肠道修复

——肠道微生物与环境污染物

Gut Remediation Gut Microbiota and Environmental Pollutants

主　编　李祥锴　刘　璞

副主编　韩华雯

编　委　云　慧　毛春兰　令桢民　葛　斌
王　星　杨　祺　姜玉超　范競文
邓杰芳　胡鑫玉　胡梓剑

兰州大学出版社
LANZHOU UNIVERSITY PRESS

图书在版编目（CIP）数据

肠道修复 ：肠道微生物与环境污染物 / 李祥锴，刘璞主编. -- 兰州 ：兰州大学出版社，2024.4

ISBN 978-7-311-06608-6

Ⅰ. ①肠… Ⅱ. ①李… ②刘… Ⅲ. ①肠道微生物－研究 Ⅳ. ①Q939

中国国家版本馆CIP数据核字(2024)第023304号

责任编辑 包秀娟 熊 芳

封面设计 倪德龙

书　　名	肠道修复——肠道微生物与环境污染物
作　　者	李祥锴 刘 璞 主编
出版发行	兰州大学出版社 （地址：兰州市天水南路222号 730000）
电　　话	0931-8912613（总编办公室） 0931-8617156（营销中心）
网　　址	http://press.lzu.edu.cn
电子信箱	press@lzu.edu.cn
印　　刷	西安日报社印务中心
开　　本	710 mm×1020 mm 1/16
印　　张	14.25
字　　数	297千
版　　次	2024年4月第1版
印　　次	2024年4月第1次印刷
书　　号	ISBN 978-7-311-06608-6
定　　价	56.00元

序 言

喜闻李祥锴教授新书《肠道修复——肠道微生物与环境污染物》即将出版，我欣然受邀为其作序。

在这个快节奏的时代，我们往往容易忽视身体的微妙变化。肠道，作为我们身体的重要组成部分，不仅关系到营养的吸收，更与我们的身体健康紧密相连。而肠道微生物，这些在肠道内默默守护健康的“小生命”，对于维持肠道的平衡起着至关重要的作用。

环境污染中的重金属、抗生素、农药残留等食源性污染物通过食物链悄然侵入我们的身体，导致肠道微生物群落失衡。肠菌生态失衡不仅影响着我们的消化与吸收功能，还会诱发一系列慢性疾病，如肠道炎症、自身免疫性疾病等。环境污染的后果，不知不觉重伤了人类自身。

李祥锴教授从事微生物学研究数十年，近年来在“西北益生菌资源高效开发与利用”领域取得了显著成果。他独具慧眼，统筹环境修复和肠道修复，创造性地利用“肠道修复”解决生物体内食源性污染物或有毒有害代谢物的积累问题。他相继开发的“浆水酸奶”“浆水百合”等功能性系列产品，为人们维护肠道健康提供了更多的选择措施和可行方案。

这本《肠道修复——肠道微生物与环境污染物》从环境污染的现状出发，深入剖析了肠道菌群与污染物的相互作用机制，阐述了益生菌在肠道修复中的广阔应用前景。同时，本书还结合丰富的案例和实践经验，探讨了益生菌在产业化应用中的发展目标及存在的问题，旨在帮助读者更好地理解环境健康和肠道健康的重要性，并学会如何在日常生活中保护自己的肠道。

在此，我要向所有为这本书付出努力的作者和编辑人员表示衷心的感谢。正是他们的辛勤工作，使得这本书能够呈现在读者面前，为读者带来一场关于肠道健康的独特阅读体验。

最后，我希望这本书能够激发更多人对肠道健康和环境健康的关注和思考，共同为构建更健康的生活方式而努力。让我们携手并进，保护好自然环境，守护好我们的肠道内环境，让人类更健康、更快乐、更长久地生活在这个蓝色的星球上！

中国科学院微生物研究所　研究员

山东大学　特聘教授

2024年4月1日

前 言

在当今的医学和健康领域，肠道健康已经成为一个热门的话题。越来越多的研究表明，肠道不仅仅是消化食物的器官，它还与人体的整体健康状况紧密相关。肠道菌群，这个由数万亿微生物组成的复杂生态系统，已被证明在许多生理和病理过程中发挥着关键作用。从免疫功能的实现、营养物质的吸收，到情绪和认知的表达，肠道菌群都扮演着不可或缺的角色。

“肠道修复”这一概念，正是基于上述科学认识而提出的。它指的是通过调节肠道菌群的平衡，来预防或缓解各种疾病，从而达到提高生活质量和延长寿命的目的。然而，随着工业化和现代化进程的加快，环境污染已成为一个全球性的问题。这些污染物不仅对人类的外部环境造成了破坏，还直接或间接地影响人体肠道菌群的稳定和多样性，对人体的内部环境即肠道的健康构成了威胁。

《肠道修复——肠道微生物与环境污染物》深入探讨了环境污染与肠道健康之间的关系，揭示了污染物对肠道菌群平衡的影响，总结了修复和维护肠道健康的科学方法。在这本书中，读者将看到关于肠道修复的最新研究成果、实用建议和策略，同时学会如何保护和恢复肠道健康。

人类生活在一个复杂多变的世界中，也面临着许多健康挑战。这些挑战，一些是自然演化形成的，还有一些是受限于人类认知而自己制造的。人类应该坚信，只要掌握了正确的知识和方法，就能够应对这些挑战，走向一个更加健康和美好的未来。愿这本书能够给读者提供一些帮助和指引，成为读者健康之旅的得力伴侣。

编　者

2023年10月24日

目 录

第1章 环境污染现状与治理问题

2021年4月22日，国家主席习近平在北京以视频方式出席领导人气候峰会，并发表题为《共同构建人与自然生命共同体》的重要讲话，他指出："人类进入工业文明时代以来，在创造巨大物质财富的同时，也加速了对自然资源的攫取，打破了地球生态系统平衡，人与自然深层次矛盾日益显现。""我们要像保护眼睛一样保护自然和生态环境，推动形成人与自然和谐共生新格局。"

1.1 环境污染

环境污染是指人类直接或间接地向环境排放超过其自净能力的物质或能量，从而使环境的质量降低，对人类的生存与发展、生态系统和财产造成不利影响的现象。随着工业化进程加速，环境污染已经成为全球关注的焦点。过度的污染不仅会对自然环境和生态系统产生深远影响，还会给人类健康和社会福祉带来严重威胁。随着科学技术的发展和人民生活水平的提高，环境污染也在增加，特别是在发展中国家。环境污染问题越来越成为世界各个国家的共同问题之一。

1.1.1 历史与发展

自从人类进入工业革命以来，无论是发达国家还是发展中国家都走了以牺牲环境换取经济发展的道路。人类的各种工业活动、农牧业活动导致环境污染以一种不可预料的态势加速发展[1]。如何缓解环境污染，并实现减少污染的目标，已成为人们密切关注的问题[2]。环境污染在人类历史上一直存在，但在过去污染主要来自重金属和生物毒素。例如，在18世纪和19世纪，工业工人使用硝酸汞处理动物皮毛用来生产帽子。当时，硝酸汞的危害还没有得到很好的认识，工人们患上了各种各样的身体和精神疾病，包括语言障碍、颤抖、情绪问题和幻觉。1817年，德国化学家弗里德里希·斯特罗梅耶发现了镉，这种重金属由于可以产生鲜艳的黄色、橙色和红色，而被广泛用于绘画。画家们在使用含镉颜料过程中吸入了有毒的镉粉末，这给他们的健康带来了巨大危害。在19世纪，人们使用有毒金属砷制作卧室的壁纸，这种卧室壁纸会导致人们生病，甚至死亡，尤其是对幼儿的危害巨大。自大约1万年前农业和谷物储存开始以来，产毒真菌产生的毒素带来的危害一直是人类面临的严重问题。最早有文献记载的毒素病可能是黑麦中的麦角菌引起的麦角菌病，这种病已经存在2000多年了。在过去的1000年里，这类病导致了成千上万人的死亡，主要发生在欧洲。在日本，一种自17世纪以来就被称为急性心脏脚气病的疾病就是由一种来自柠檬青霉的真菌毒素引起的。而伴随着现代城市化和工业化与先进的生产技术相结合，新的环境污染物也不断进入我们的日常生活，在过去的几十年里，新出现的污染物，如药用活性化合物、个人护理产品、人工甜味剂和内分泌干扰化学品等越来越受到人们的关注，因为它们在环境中无处不在，并有可能对生态环境造成不良影响[3-7]。根据Chen和Ye在2014年的报道，中国有300多万公顷土地因被污染而无法正常进行农业活动[8]。土壤沉积物中的重金属和有机污染物是两种主要环境污染物[9, 10]，这些污染物对人体健康和生态环境已经构成巨大的威胁。为了减少这些污染物带来的健康风险，必须了解各类污染物的性质。

1.1.2 主要环境污染物

（1）重金属

随着近现代工业化和城市化的迅速发展，重金属污染已经成为主要的环境问题，由于重金属持久性、普遍性和高毒性地存在，重金属污染已经严重影响了生态系统[11, 12]。影响人类健康的主要有毒重金属如锌、铜、镍、锡、铅、镉、汞、铬和砷等已经在环境中广泛传播[13]。

重金属是不可生物降解的，人类会在不经意间通过食用各种各样的食物将重金属积累到身体中[14]。与此同时，有毒金属元素会从各种自然和人为源头被释放到水中。据报道，全球多地地表水中发现铬、锰、铁、钴、镍、砷、镉等重金属的平均浓度远高于饮用水的最大允许值[15, 16]。地球上大约40%的湖泊和河流已经受到重金属污染[17]。这些不能被生物降解的重金属会随着时间的推移在各类生物体中积累，且在生物体中的浓度会持续增加，并可以直接或间接地通过生物放大作用影响各种生物[18]。有报道表明，在环境中即使低浓度的重金属也会造成人体皮肤、食道、胃、肺、肝、肾和前列腺等多个器官的损伤，更为严重的是还会损伤中枢神经系统[15, 18]。重金属在各类器官中的不断积累，将导致氧化损伤、内分泌紊乱和免疫系统抑制，严重影响生物的生存和生长[19]。重金属会对生物产生严重的危害，因此重金属污染在全球范围内受到了广泛的关注。Hu等在2020年收集了2008—2018年在中国发表的1200多项重金属污染的研究数据，并在此基础上，评价了中国重金属污染状况及相关健康风险，探讨了中国重金属污染的空间格局及潜在主导因素[20]。如图1-1为中国重金属污染在各个省域的分布情况。

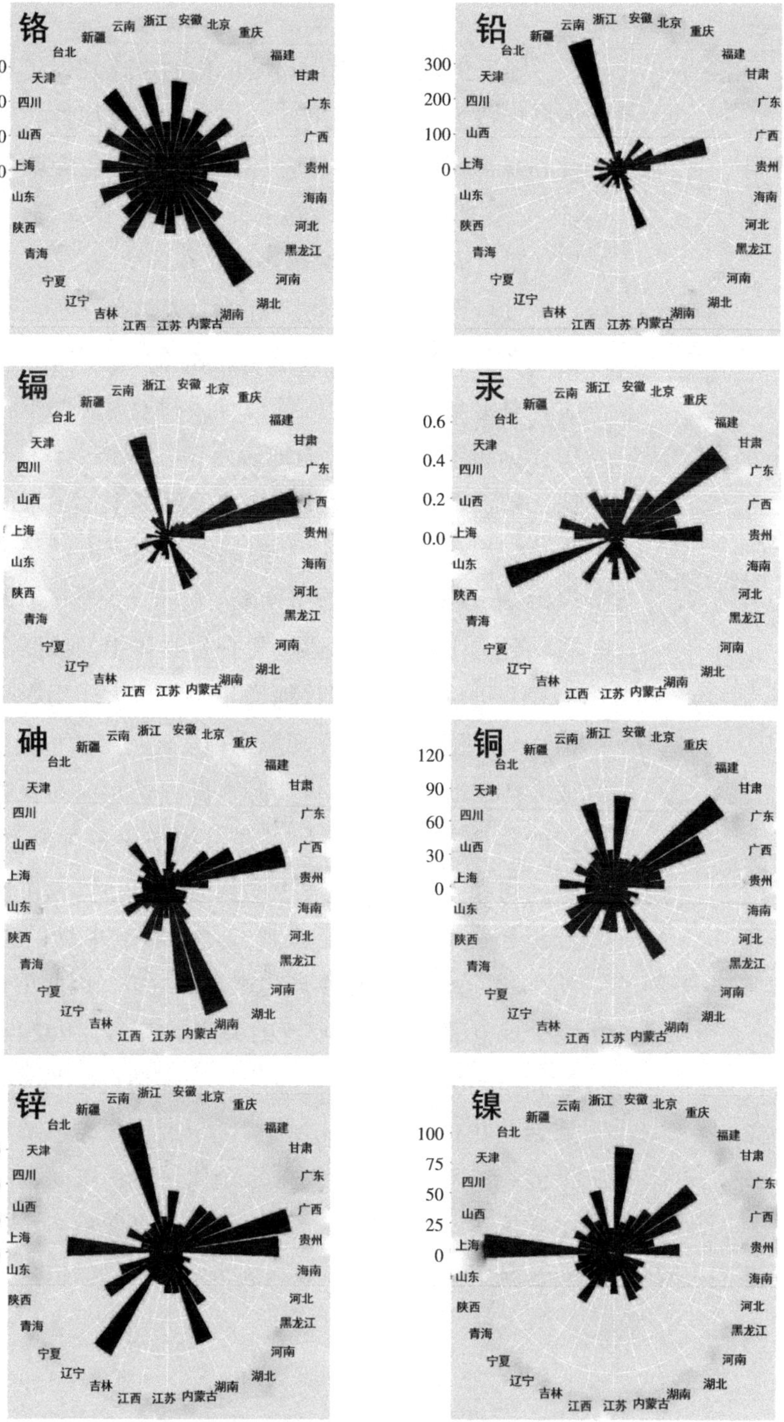

图 1-1　中国重金属污染省域分布情况(单位:mg/kg)[20]

为了更好地应对重金属污染对人体健康的影响，下面将介绍几类环境中广泛存在的重金属对人类健康的危害。

镉通常以杂质的形式存在于锌矿或铅矿中，是冶金工业、采矿工业等一些化学工业的副产物[21]。中国是世界最大的镉生产国和消费国之一，大量有毒的镉已经被释放到了环境中。镉具有持久性和毒性，镉污染对生态环境和公众健康构成了严重威胁[22，23]。长期接触镉会对人体的肾脏和骨骼产生不利影响，肾脏是镉聚集的主要部位[24]。镉会降低鱼的生长速度，破坏免疫系统和血液学参数，导致氧化应激、组织学损伤、行为改变和DNA损伤[25]。镉是应用最广泛的重金属之一，被应用在一系列的工业中，如电池、油漆、电镀、化肥和塑料业等[26]。镉的广泛应用导致其大量暴露，而镉暴露会给人体健康造成巨大危害，如导致肝毒性、肾功能障碍、心血管疾病、骨质疏松、免疫功能低下和肿瘤等[27，28]。

砷是一种天然污染物，也因人为使用而被释放到环境中，主要来源于采矿和冶炼工业。砷的暴露会影响多种生物系统，并导致皮肤癌、肺癌、膀胱癌和其他组织癌[29]。

铬是颜料生产、金属电镀和皮革制造等行业中最常用的重金属之一，主要以两种形式存在：三价铬和六价铬[30]。六价铬因其具有致突变性和致癌性而引起人们的极大关注[29]。接触六价铬会引发肝毒性、肾损害、炎症和肠道微生物群失调[31]。铬污染主要来源于各种人类活动，如橡胶工业和染料工业生产和污水处理都会造成铬污染[32]。铬化合物可对人体造成的影响包括导致癌症、肝毒性、肾损害，以及影响中枢神经系统，甚至可能导致死亡[33]。

汞是环境中最危险的四种重金属之一，主要来源于火山爆发和制革、采矿、电镀等工业化学品或废弃电器产品的加工过程。汞中毒的症状包括肌肉无力、协调性差、手脚麻木、皮疹、焦虑等[34]。所有形式的汞在高剂量时都会对人体健康产生不利影响，如引发牙龈炎、胃肠功能障碍和急性肝毒性[35]。在20世纪50—60年代的日本，汞污染海水和鱼类，导致熊本县数千人中毒而发生的水俣病，就是食用汞污染鱼类的后果[36]。

铅可导致人类发生不可逆的神经损伤，并对肾脏、心血管和消化系统造成损害[37, 38]。环境中的铅主要来源于人造产品，如含铅汽油和油漆、铅酸电池和柴油废气，环境中的铅最终通过自然循环排入水生生态系统[39]，这种循环将会带来严重的环境风险。表1-1为主要重金属来源及其对人类的危害。

表1-1　主要重金属来源及其对人类的危害

重金属种类	来源	对人类的危害
砷[40]	采矿、冶炼	皮肤病、癌症、心血管疾病
镉[40,41]	冶金、采矿和化学工业	心血管疾病、骨质疏松、免疫功能低下、癌症
钴[42]	电镀、冶金、制革、油漆等工业	低血压、骨骼畸形、瘫痪
铬[43,44]	污泥、固体废物、橡胶工业	癌症、肝毒性、肾损害、中枢神经系统病变
铜[45]	合金生产、电镀和采矿工业	睡眠障碍、肝脏并发症
汞[40,44]	火山爆发、制革、采矿、电镀	牙龈炎、胃肠功能障碍、急性肝毒性
锰[43]	石油、采矿、杀虫剂	神经损伤、肝性脑病
镍[41,43]	工业废水	呼吸障碍、癌症、皮肤病
铅[41]	城市污水、工业废物	中枢神经系统损伤、高血压
锌[43,45]	电镀、采矿	焦虑、睡眠障碍

（2）*生物毒素*

生物毒素是一类由生物自身产生的天然毒素，包括植物、动物和微生物等产生的对其他生物物种有毒害作用的各种化学物质[46]。这些生物毒素在环境中的循环将会影响人类的健康。在这里，将介绍一些在食品和环境中存在的最重要的生物毒素。

真菌毒素是由真菌在代谢过程中产生的一些有毒次生代谢产物[47]。在诸多不同种类的真菌毒素中，黄曲霉毒素对人类构成的威胁是最严重的。黄曲霉毒素是由黄曲霉和寄生曲霉产生的诱变剂和强致癌物。黄曲霉毒素存在于腐烂的食物中，如木薯、花生、大米、芝麻、葵花籽、玉米、坚果、小麦等。黄曲霉毒素暴露对儿童的影响最大，会导致其生长发育迟缓[48]。真菌毒素会导致人和动物的肠道病变，包括使肠上皮坏死[49]，扰乱肠道屏障功能，破坏免疫反应，并通过影响生长激素分泌直接影响食物消耗[50]。一些真菌毒素在肠道菌群中诱导DNA修饰和突变[51]。蓝藻毒素是由出现在湖泊、池塘、河流和其他地表水中的蓝藻细菌产生的毒素。蓝藻菌在营养丰富的条件下会引发有害藻华[52]。有研究表明蓝绿藻能够释放毒素，毒素进入人体后会损害皮肤、肝脏和神经系统[53，54]。

（3）*新兴污染物*

新兴污染物是指新近发现的，并且尚未纳入系统管理以及现有的管理措施不足以有效防控其风险的污染物，这类污染物对人体健康和生态环境存在风险。新兴污染物已在各类环境中被频繁检测出[55]。随着工业化进程的加快，大量化工产品在为生产生活提供更加丰富、优质服务的同时，也随之带来越来越多的新兴污染物。目前，国际上尚未就新兴污染物的分类达成共识，通常而言，内分泌干扰物、药品与个人护理用品、全氟化合物、溴代阻燃剂、饮用水消毒副产物、微塑料等都属于该范畴的污染物。这些污染物通过多种途径被引入，包括直接排放城市废水、工业废水、污水处理厂排水，以及畜牧业粪肥废物等[56]。污水处理厂是新兴污染物的主要来源之一，污水处理厂的原始进水和处理出水中经常检测到新兴污染物，污水处理厂在不断地向环境中排放新兴污染物。持续排放和长期暴露于微量水平的这些新兴污染物下也可能对人类健康构成威胁[57]。令人更为担忧的问题是，新兴污染物在环境中的广泛存在将会进一步导致抗生素耐药细菌和基因的出现，从而降低人类使用抗生素治疗疾病的潜力[58，59]。

在这里，我们将重点介绍内分泌干扰物、药品与个人护理用品这两类具有环境污染风险的新兴污染物对人类和其他生物健康的影响。内分泌干扰物是指环境中存在的能干扰人类或野生动物内分泌系统诸环节并导致异常效应的物质，主要包括农药、食品添加剂等。药品与个人护理用品大多数极性强、易溶于水、挥发性较弱，是一类微妙的、潜在的、有积累影响作用的环境污染物，其中抗生素由于能引起微生物的选择性压力和抗药病原菌的选择性存活而受到广泛关注。

1）农药

随着现代化农业的快速发展，农业活动中农药的使用量正在增加，虽然近年来已经开发出相对安全的农药，但是有害的农药仍然在一些地区被广泛使用。农药根据其化学结构主要可分为四大类：有机氯、有机磷、氨基甲酸酯和拟除虫菊酯[60]。根据其作用可分为除草剂、杀虫剂、杀菌剂和植物生长促进剂[61]。化学农药进入人体会给人体带来各种不良反应，如导致皮肤过敏、呼吸道黏膜受损、消化道异常、神经毒性反应等[62，63]。低浓度的高毒性农药会刺激自由基的产生，导致DNA损伤、脂质过氧化和细胞死亡，从而对内分泌系统、皮肤和神经系统产生长期的不良影响[64，65]。

有机氯农药是一类在化学工业和农业中广泛使用的合成农药。有机氯以其高毒性、慢降解和生物蓄积等特点而闻名。有机氯中毒主要会出现以中枢神经系统机能紊乱为特征的兴奋、痉挛、麻痹等典型症状。环二烯抑制钙离子流入和相关酶的活性，导致神经递质的释放[66]。流行病学研究表明，帕金森病与有机氯污染物之间存在病因关系，这类化学物质可能对人类健康造成巨大的危害[67]。

有机磷农药是一类用在农业活动中的杀虫剂[68]。有机磷农药在环境中的污染来源十分广泛，主要来源于林业、农业、医药等领域。自20世纪70年代以来，世界上各国禁用有机氯农药后，有机磷农药就成为农业生产活动中主要的害虫防治化学品之一。有机磷农药在土壤、水体环境中的残留范围广、残留量高、残留时间久。有机磷农药经过生物累积、生物放大等途径影响人体健康，在临床表现出一系列生理毒性和神经毒性等。

氨基甲酸酯类农药是一类可干扰昆虫内分泌系统的化学农药[63]。氨基甲酸酯类农药因其可有效地抑制乙酰胆碱酯酶的活性，被广泛用作杀虫剂[69]。氨基甲酸酯类农药通过影响昆虫线粒体和细胞代谢等功能来影响昆虫的生殖[70]。氨基甲酸酯同样也可以抑制人类中枢和外周神经系统中的乙酰胆碱酯酶活性[71]，引起神经递质乙酰胆碱在体内积聚，进而导致人体急性中毒[72]。不当和过量使用氨基甲酸酯会在环境中留下大量的氨基甲酸酯残留物。长期摄入残留水平超过安全标准的食物可能会导致人体慢性中毒，甚至带来癌症风险[73]。

拟除虫菊酯类化合物是一类从菊花中分离出来的天然化合物[74]。目前，拟除虫菊酯已被广泛应用于农业、林业和纺织工业，且其在医学和兽医学中被用于甲壳类寄生虫病的防治[75]。拟除虫菊酯是常用的一种杀虫剂，因为其对昆虫具有高毒性，并且可快速生物降解。拟除虫菊酯是脂溶性的，因此与皮肤、消化道和呼吸道的任何接触都会导致其被渗透到体内[76]。与其他人群相比，儿童和孕妇面临的拟除虫菊酯渗入风险更高[77]。

2）食品添加剂

长期以来，人们一直使用人工甜味剂增强食物风味。人工甜味剂取代蔗糖的同时降低了食物卡路里和人们患龋齿的风险[78]，但它们的安全问题一直存在争议。由于它们在环境中持久性存在，对生物会有潜在的不利影响，被认为是一类新的污染物[79]。人工甜味剂在食品添加剂和个人护理产品中被用作糖的替代品。迄今为止，关于人工甜味剂在环境相关浓度下对生物和人类健康的毒性报告很少，但是长期使用人工甜味剂可能对生态系统安全带来越来越大的风险[80]。有研究显示，食用人工甜味剂会增加肥胖、高血压、糖尿病和心脏病的风险[81]。人工甜味剂对包括糖尿病患者、儿童、孕妇、育龄妇女、哺乳期妇女等易感人群有潜在的有害影响，需要对这些易感人群进行更多的研究来减少人工甜味剂对其的危害[82]。三氯蔗糖也是常用的人工甜味剂，其甜度为蔗糖的600倍[83]。一项动物研究表明，三氯蔗糖可能导致有益肠道菌群的减少，这可能造成有害的健康影响[84]。

三聚氰胺是氰胺的三聚体，具有1，3，5-三嗪骨架[85]。有动物研究表明，高剂量三聚氰胺可引起结石和膀胱癌[86，87]。2008年，中国爆发了牛奶和婴儿配方奶粉以及其他食品原料掺入三聚氰胺的事件。添加三聚氰胺可增加牛奶中的氮含量，使牛奶的蛋白质含量变高。在这起事件中，中国约有30万名受害者，其中6名婴儿死于肾结石和其他肾脏损伤，超过5万名婴儿住院。2008年发生在中国的三聚氰胺事件揭示出三聚氰胺在过量食用时会产生毒性[88]。

防腐剂是一种添加到食品和饮料等产品中，以减缓微生物生长或化学变化，从而防止食品腐败的化学物质。防腐剂含有一些有毒化学物质，如硝酸盐和含苯物质。研究表明，硝酸盐可诱发动物胰腺肿瘤[89]。苯甲酸盐被用作酱油和软饮料的食品防腐剂，在抗坏血酸和过渡金属催化剂的作用下，苯甲酸会脱羧生成苯[89]，而苯是已知的致癌物质。

3）抗生素

抗生素是一类可以治疗细菌感染引起的疾病的抗菌物质。自从1928年亚历山大·弗莱明发现第一种抗生素——青霉素到如今，人类已经发现和人工合成了多种抗生素，并将其用于疾病治疗。抗生素通过控制微生物感染以挽救生命，在医疗领域中被广泛应用[90]。抗生素也被用作杀虫剂和饲料添加剂，用来提高农业和畜牧业的产量[91，92]。然而，大量使用抗生素导致抗生素不能被有效利用，进而被大量释放进入环境。有报道表明在世界土壤环境中抗生素污染广泛分布[93]。更为严重的是，大量抗生素暴露在环境中会诱导新的抗生素耐药细菌与基因的出现，给人类社会与全球环境带来了新的严重的威胁[94，95]。据统计，在美国和欧洲每年至少有2万多人死于抗生素耐药性[96]。到2050年，全球每年因为抗生素相关危害造成的死亡人数将超过1000万，相关医疗费用将超过100万亿美元[97]。大量的抗生素通过人们的日常生活、医院排水、污水处理厂、养殖业被释放到环境中[98-100]。人类活动如农业生产中使用抗生素会导致土壤中外来抗生素抗性细菌的存在[101，102]。畜牧业产生的粪污废水中也检测到大量抗生素残留[103]。制药工业的流出物被认为是造成水体抗生素污染的重要因素[104]。多种抗生素存在于水环境中也会对水生生物构成生态风险，并会促进抗生素抗性基因和水生生态系统中的抗抗生素细菌的出现[90，100，105]。抗生素的环境暴露对人类健康带来了极大的风险。

在此，将介绍一些较为常见的抗生素及其对人类健康的影响。

①克拉霉素可用于治疗细菌性疾病，如肺炎、链球菌性咽喉炎、幽门螺杆菌感染、皮肤感染等。克拉霉素

最常见的副作用是引起胃肠道不适，如恶心、腹痛、腹泻和呕吐[106]。

②甲硝唑是一种抗生素和抗原性动物药物，对盆腔炎、心内膜炎、细菌性阴道炎有效。其副作用包括恶心、腹泻、体重减轻、腹痛、呕吐、头痛、头晕和口腔金属味[107]。

③环丙沙星是一种广谱抗生素，可用于治疗细菌感染引起的疾病，包括骨和关节感染、腹腔感染、皮肤感染、呼吸道感染、尿路感染、感染性腹泻、伤寒等。常见的副作用有恶心、呕吐、腹泻和皮疹[108]。

④克林霉素是一种可用于治疗多种细菌感染的抗生素。它对骨或关节感染、盆腔炎、链球菌性咽炎、肺炎、中耳感染和心内膜炎有效，也对革兰氏阳性菌、厌氧菌和支原体有效，但对革兰氏阴性菌无效。常见的副作用包括腹泻、恶心、呕吐、腹痛或痉挛或皮疹[109]。

⑤氨苄西林是β–内酰胺类抗生素，其活性与阿莫西林相似。氨苄西林的毒性比大多数抗生素都小，副作用包括血管性水肿和过敏反应。氨苄西林因可能引起过敏反应，对青霉素类过敏的患者禁用。过敏反应包括多形性红斑、剥脱性皮炎、皮疹、荨麻疹，以及体内红细胞与白细胞暂时减少等[110]。

1.1.3 环境污染的发生途径

环境中的各种污染物，如重金属、新兴污染物、有毒气体等伴随着人类的日常活动进入人体。这些污染物在人体内积累会带来严重的健康风险，其主要在土壤、水体和大气间传播，如图1–2所示。下面将介绍污染的主要发生途径，借此提高公众的环境保护意识。

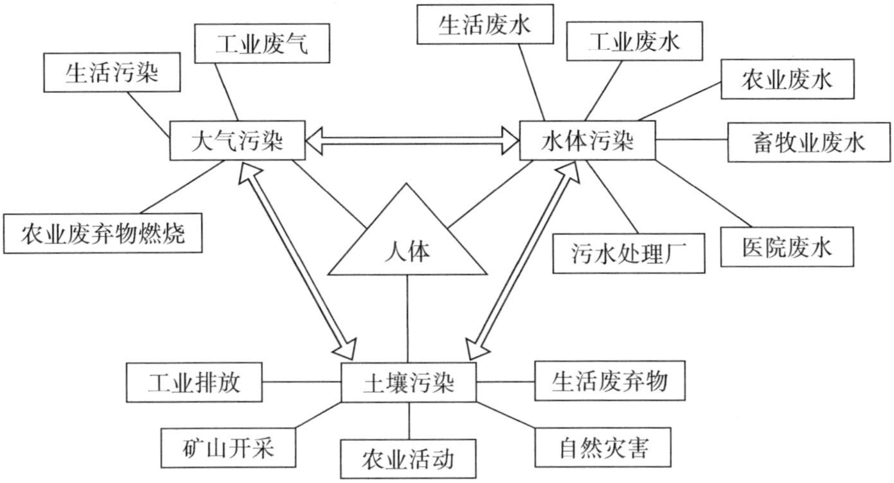

图1-2 环境污染主要发生途径

（1）土壤

土壤是与人类生活密切相关的重要物质资源，也是环境间物质循环的重要场所[111]。土壤污染对生态环境和公众健康的危害极大，已成为世界性的重大问题[112]。由于过去几十年的快速工业化和人口过剩，环境中污染物大幅积蓄。这些污染物过量进入环境会导致人类和环境的健康问题，即使浓度很低，也会影响土壤、人类、植物和动物。水污染也会导致土壤污染，反之亦然。自然水体中的污染物可以从水体渗透到土壤中，这将会对人类健康和生态环境造成严重威胁。大量污染物通过自然循环和人为活动在土壤环境中传播。各种人类活动，如废物处理、废水处理、汽车尾气排放、矿山开采，以及农牧业和工业等都有利于污染物在土壤中积累。人们为了提高生产效率，不可避免地会在农业生产中使用大量的化肥和农药，而使用过程中却不按照相关标准执行，使得大量污染物随之进入土壤，后在农作物中积累，最终通过食物链进入人体，进而影响人们的身体健康和生命安全。自然恢复土壤污染需要很长的时间，即使采取合理的人为修复方式修复也无法使其彻底恢复，并且修复过程中付出的经济代价远大于采取防治措施的代价。

（2）水体

人类各种活动产生的各类污染物若未经处理或处理不完全就被排放进入水体，则会造成水体污染，使得水资源的利用价值降低甚至丧失。水体污染的主要来源有以下几方面：

①工业污染。各类工业生产过程中产生的废水是水体污染的最主要来源，如冶金、电镀、造纸、印染、制革等。

②生活污染。人们生活中排放的各种污水，如洗涤衣物、生活垃圾废弃物处理产生的污水。

③农牧业污染。农牧业污染主要来源于农药、化肥和兽用药品的不正确使用。为了提高农作物产量而长期施用过量化肥和农药，导致大量化肥与农药在农田系统中残留。

④医院污水。医院大量使用各类药物和消毒剂等，如果使用浓度过大且处理不充分就会造成水源污染。

⑤污水处理厂。污水处理厂是水体污染的主要发生途径。通过污水处理厂向水生系统输送微污染物的主要废水类型包括生活废水、医院废水、工业废水和农牧业废水，而水生环境中微污染物的大量暴露往往会带来诸多负面影响，包括短期和长期毒性、内分泌干扰效应和微生物的抗生素耐药性等[113]。

（3）大气

随着工业化及城市化的快速发展，大气污染已成为世界各国不得不面对的重大环境问题。在世界工业革命时期，发生过伦敦烟雾、美国光化学烟雾等大气污染引发的危害居民生命健康的事件[114]。有研究证实空气污染物会通过呼吸道进入人体，对人体免疫系统产生影响[115]。

大气污染的来源以人为活动为主，主要来源于工业废气、生活污染、农业废弃物燃烧等。

1.1.4 环境污染的危害

我们生活在一个被化学物质包围的世界，这就需要我们了解化学物质对人类的危害，并评估它们引起的环境污染对人类健康和环境安全的潜在风险。以下列举20世纪世界上重大的污染事件，希望借此提高公众对污染的认识，以减少环境污染的进一步恶化。

①水俣病事件。该事件的起因是位于日本熊本县水俣镇的一家肥料生产公司排放的废水中含有大量的重金属汞，这些废水未经有效处理就被排入了大海。这些未经有效处理的汞首先在海水和鱼类中富集，后经过食物链进入人体使得人中毒。据日本环境厅1991年公布的数据，此次事件中的中毒病人有2248人，其中1004人死亡[36]。

②骨痛病事件。骨痛病事件是由于日本富山县当地人从事的矿山开采活动排放的废水中含有大量的镉，导致大量的重金属镉在当地河流中积累。而当地的人食用含镉河水浇灌作物，加上长期饮用高镉含量的水，导致其骨骼严重畸形、骨脆易折，且出现骨痛等[116]。

③日本米糠油事件。日本米糠油事件发生在20世纪70年代，是因为当地人食用了含有多氯联苯的米糠油，最终致使他们出现眼皮发肿、手掌出汗、全身肌肉疼痛、咳嗽不止、肝功能下降等症状[117]。

④印度博帕尔事件。在20世纪90年代，美国联合碳化公司开设在印度博帕尔市的一个农药制造厂因管理制度混乱与员工操作不当，导致剧毒物质甲基异氰酸酯的高压储存罐发生爆炸。此次事件波及范围达40 km^2，造成近两万人死亡，20多万人受害，5万人因此失明，同时也致使孕妇流产或产下死婴，数千头牲畜被毒死[118]。

⑤剧毒物污染莱茵河事件。位于瑞士巴塞尔市的一家大型化工厂的储存仓库发生火灾，致使近30吨剧毒的磷化物、硫化物与含有重金属的一系列化工产品进入莱茵河流域。该事件导致莱茵河流域生态环境被严重破坏，河流中的鱼被大量毒死，河岸两侧的井水不能饮用，同时有毒物大量沉积在河底，莱茵河流域在之后的20年也因此“死亡”[119]。

从《中国21世纪议程》的发表到科学发展观的提出，再到生态文明建设的推行，中国的可持续发展之路正朝着更深、更强、更全面的方向前进[120]。然而，随着经济的快速发展，中国也面临着日益严峻的环境和资源问题，发展的不可持续问题也日益突出。中国不希望在发展进程中重蹈一些西方国家“污染后治理”的覆辙，这就需要全国人民付出更多的努力，走出一条具有中国特色的环境保护道路。

1.2 环境保护

随着中国经济的发展，有效利用能源，减少环境污染，降低安全生产事故频次，防止突发环境事件，确保生命安全的重要性日益凸显。制定并执行环保政策和措施，在保护环境的同时改善人民的生活质量，已经成为中国民生工程的关注点。保护环境不仅关乎人们的生存环境，也影响着经济发展。

1.2.1 环境保护的提出与发展历史

环境保护这一概念是人类的生产活动迅速发展导致的环境污染问题过于严重时，引起世界各个国家的重视而产生的。环境保护涉及的范围广、综合性强，它涉及社会科学和自然科学的许多领域[121]。各国政府利用国家法律法规对环境污染进行约束，从而使环境保护逐步引起全社会的重视。

20世纪60年代，美国著名生物学家蕾切尔·卡逊出版了一本名为《寂静的春天》的书，在书中他向人们阐述了农药杀虫剂对环境的污染风险。由于该书对环境恶化的警示，美国首先开始对剧毒杀虫剂污染进行了防治处理，先成立了环境保护局，之后又相继出台了关于生产和使用剧毒杀虫剂的规范和法律。然而“环境保护”这一术语被广泛地采用是在1972年召开的联合国人类环境会议以后。

中国的环境保护事业是从1972年起步的，之前也存在“综合利用”工业废物方针、“三废”处理和回收利用的概念，直到20世纪70年代改用“环境保护”这一科学的概念。根据《中华人民共和国环境保护法》的相关规定，环境保护的内容主要包括保护自然环境与防治污染和其他公害两个方面。

环境问题是中国在新世纪以来面临的一大严峻挑战，自然环境的稳定是一个国家实现经济长期稳定增长和可持续发展的基本前提。环境问题解决得好坏关系中国的国家安全、国际形象、广大人民群众的根本利益，以及全面建成小康社会的实现。时任国务院总理的温家宝在第十二届全国人民代表大会作《政府工作报告》时也曾指出，要顺应人民群众对美好生活环境的期待，大力加强生态文明建设和环境保护，生态环境关系人民福祉，关乎子孙后代和民族未来，要坚持节约资源和保护环境的基本国策，着力推进绿色发展、循环发展、低碳发展，要加快调整经济结构和布局，抓紧完善标准、制度和法规体系，采取切实的防治污染措施，促进生产方式和生活方式的转变，下决心解决好关系群众切身利益的大气、水、土壤等突出环境污染问题，改善环境质量，维护人民健康，用实际行动让人民看到希望。保护环境是中国长期稳定发展的根本利益和基本目标之一，实现可持续发展依然是中国面临的严峻挑战[122, 123]。

1.2.2 环境保护政策

自1978年改革开放以来，中国经济社会飞速发展，取得了举世瞩目的成就，中国已成为仅次于美国的世界第二大经济体[124]。但是，在经济快速增长的同时，也造成了环境的恶化，带来了一系列的环境问题，如自然资源耗尽、生态污染、生态系统退化等。这些环境问题对中国人的生命安全以及中国的可持续发展构成了巨大的威胁[125]。以空气污染为例，2013年中国多地出现空气质量指数“爆表”现象，全国雾霾天气频发，1/3的城市出现雾霾问题，这严重威胁人民健康、制约经济发展[126]。此外，经济发展和人口快速增长导致水污染问题越来越严重[127]，工业活动造成的土壤污染也日益严重和普遍。为此，中国政府先后制定了一系列政策，以实现环境的可持续发展[128]。随着经济的发展和公众环保意识的提高，中国环境保护的核心原则从以前简单的“三废”治理转向生态文明建设[129]。近年来，中国依据自然界生态循环规律，陆续提出了可持续发展、科学发展观等一系列绿色发展理念，党的十八大把生态文明建设纳入“五位一体”的总体布局，党的十九大更是把美丽中国作为建设社会主义现代化强国的重要目标。因此，环境保护的地位和作用得到了极大的提高，中国环保事业也踏上了具有中国特色的环境保护道路[130，131]。

尽管中国政府已经开展了许多环境和生态项目来解决环境问题，如低碳减排和开发可再生资源等，但随着经济的发展，新的环境问题仍然不可避免，环境保护制度体系还存在较大的完善空间[132]。以下是中国环境保护政策发展四个阶段的详细分析。

（1）探索起步阶段（1972—1991年）

中华人民共和国成立之初，由于经济发展滞后，造成的环境污染与生态破坏也只发生在部分地区[133]。1972年之前中国并未制定和实施系统的环境保护政策。1972年，斯德哥尔摩会议通过了《人类环境宣言》，并提议将每年的6月5日定为“世界环境日”，这次会议是中国政府重返联合国后参加的第一个国际会议。在这次会议上，中国政府开始认识到中国存在的环境问题，并开启了中国人民环境保护意识觉醒的伟大新征程[134]。1973年，第一届全国环境保护会议通过了《关于保护和改善环境的若干规定（试行草案）》，这是中国的第一个综合性环境保护法规，标志着中国严格意义上环境保护工作的开始[135]。1978年，中华人民共和国第五届全国人民代表大会第一次会议通过了经修订的《中华人民共和国宪法》，其总纲第十一条明确规定“国家保护环境和自然资源，防治污染和其他公害”，这标志着“环境保护”正式列入中国宪法大纲，环境保护被确定为中国的一项基本职责，中国人对环境和资源保护的重视程度提升到了一个新高度[132]。1979年9月13日，《中华人民共和国环境保护法（试行）》颁布，这是新中国成立以来颁布的第一部环境保护基本法，标志着中国环境法制化建设开始启动，是中国环境立法走向体系化的一个重要里程碑[135，136]。1983年第二次全国环境保护会议明确指出，环境保护是一项基本国策，这说明环境保护在中国经济和社会发展中具有重要作用，这一基本国策的确立对未来中国环境保护规划的实施具有深远影响[133]。另外，中国还陆续修订了《生活饮用水卫生规程》《工业企业设计卫生标准》等相关标准，强化了对人民健康的法治保障，体现了全面的环境立法开始与日益恶化的环境作斗争。

现阶段环境管理的重点是实现城市环境的综合改善和工业污染的控制[137]。初期，国家污染治理通常采取事后监督罚款的办法。但由于罚款金额不当、企业对环境政策抵触等原因，一些企业宁愿承受罚款也不愿费工夫治理污染，环境治理出现治标不治本的局面，没有从根本上有力遏制环境污染[133，138，139]。20世纪80年代到90年代初，中国出台了一系列的环境保护法律和法规，鼓励企业制定规章制度，主动参与污染治理和环境保护，环境政策向污染前控制转变。自此，中国的环境保护法规体系逐渐完善，环境保护工作开始有了较为全面的法律依据[140]。现阶段的环境保护法律法规虽然在化工和重金属行业的污染控制中是有效的，但也存在一些问题，如信息不对称造成资源浪费、企业自觉意识差等。环境治理模式也存在一些问题，例如，环境保护机构一直处于临时性、非正式和不独立的状态，环境保护工作缺乏其他利益相关方或地方当局的参与[140]。此阶段制定的环境保护指导原则是“环境保护与经济社会发展相协调”，但经济发展必须优先于环境保护，忽视了环境保护的第一要务性。总之，这一时期，中国对环境保护做出了初步探索，环境治理有了法律基础，环境污染与生态破坏的趋势得到了一定程度的缓解，环境保护事业正式起步。

（2）发展阶段（1992—2001年）

在探索起步阶段建立的良好基础之上，随着改革开放后中国工业化的高速发展，中国环境立法进入了快速发展阶段。1992年，中国参加了在巴西召开的联合国环境与发展会议（United Nations Conference on Environment and Development，UNCED），在会议“可持续发展”观念的影响下，中国制定了环境保护十大战略方针，以实现可持续发展[141]。从1993年起，中国最高立法机构全国人民代表大会不仅通过了新的环境保护法，而且还修订了许多现有的环境法。1994年，中国政府颁布了《21世纪议程》，该议程确定了21世纪人口、环境和经济发展的详细规划目标，将可持续发展纳入经济社会发展长远规划，并逐步将可持续发展理念作为环境立法的宗旨[142]。1998年大洪水过后，生态保护的紧迫性受到进一步重视，中国政府将“治污与生态保护”放在同等重要的位置，并据此制定了一系列政策，如禁止在长江和黄河流域采伐天然林，恢复西部地区的生态等，这标志着中国生态保护事业的成功转型[137]。2002年第五次全国环境保护会议召开，会议把环境保护列入政府职能，中国的环境保护工作也因此进入了一个新的历史时期。在以上重要节点的间隙，中国还不断总结环境保护经验，严格执行环境保护法，并陆续出台了《中华人民共和国节约能源法》（1997年）等重要环境保护法律[143]。

在这一阶段，随着新的环境问题的出现，政策和决策风格逐渐发生变化，环境管理由事后治理调整为事前监督的污染预防[137]。在"九五"期间（1996—2000年）国家污染治理总投资迅速增长的同时，公众开始积极参与环境保护和生物多样性保护。自此，中国逐步形成了强调环境与经济同步、协调、持续发展的可持续发展战略[144]。1999年，中华环境保护基金会（China Environmental Protection Foundation，CEPF）首次发起了一个真正的排放交易计划，从最初实行的"谁污染，谁治理"，转变为着力点源控制与浓度控制。到"九五"期间，全国污染治理普遍化，大规模的环境基础设施建设也提上日程，这体现了中国环境管理从点源治理到区域流域联防联控的转变[133]。总的来说，该阶段各项环境政策更加有机统一，整体性不断提升，但公众参与的积极性还需进一步发掘，让群众加入环境政策制定、执行的全过程更是大势所趋[138]。

（3）深化阶段（2002—2012年）

从本质上讲，环境问题是发展问题。因此，中国提出了一系列保持经济与环境平衡发展的新思路，以指导生态环境保护。第三阶段的环境保护政策在严格控制污染物排放的前提下，很大程度上保持了环境与经济的协调关系。进入21世纪后，许多新的经济概念开始出现，包括低碳发展（2009年）、生态文明建设（2012年）。这些概念为认识中国的环境保护与经济发展之间的关系提供了新的视角，也说明环境问题和经济发展促进了概念的创新。2002年《中华人民共和国清洁生产促进法》的颁布与施行，宣告中国环境管理进入全过程控制的新时期[138]。2003年10月，中共十六届三中全会通过了《中共中央关于完善社会主义市场经济体制若干问题的决定》，该决定提出“五个统筹”和坚持以人为本，树立全面、协调、可持续的发展观，这标志着环境保护进入深化发展阶段[133]。2006年，第六次全国环境保护会议提出要把环境保护摆在更加重要的战略位置，进一步加强环境保护工作，落实环境影响评价、污染物排放总量控制和环境目标责任制[133]。2009年环境保护部（现生态环境部）组建了12个环境督查中心与核辐射安全监督站，初步形成了辐射全国的环境安全管理网络[140]。截至2010年底，18个省开展了生态补偿试点，补偿方案包括矿产回收、流域环境保护和自然保护区建设[145]。2012年中国公布了更为严厉的节能减排工作方案，明确了中国节能减排的具体任务和目标，同年生态文明建设也被纳入了“五位一体”的全面建成小康社会的总体战略布局[140]。

总的来说，在这一时期，中国的环境管理体制变化主要表现为：全国人民的环保理念得到了前所未有的加强，环境管理的组织机构进一步完善，环境管理体制进入了深化发展阶段。

（4）综合改革阶段（2013年至今）

党的十八大提出，要将生态文明建设纳入中国特色社会主义事业“五位一体”总体布局，建设美丽中国，实现中华民族的永续发展，这标志着环境保护被提上了新的高度。党的十九大进一步把“坚持人与自然和谐共生”作为新时代坚持和发展中国特色社会主义的基本方略之一，强调“必须树立和践行绿水青山就是金山银山的理念，坚持节约资源和保护环境的基本国策”[146]。在以上思想的指导下，中国相继出台了一系列新的环境管理政策法规，生态文明建设得到了强有力的支持。2014

年4月24日，十二届全国人大常委会第八次会议修订了《中华人民共和国环境保护法》，该法于2015年1月1日起正式施行，被称为史上最严厉的环保法律，以此为基础，中国环保立法体系趋于成熟。新的环境保护法施行后，《中华人民共和国土壤污染防治法》(2019年)、《中华人民共和国长江保护法》(2021年) 等环境保护法相继颁布，明确提出了土壤质量、水环境质量的改善目标和分类管控措施，促进了生态环境的保护与修复的同时保障了公民的安全与健康。

党的十八大以来，环境法治建设在习近平总书记"最严法治观"的指引下，承担起保障统筹推进"五位一体"总体布局、协调推进"四个全面"战略布局的新使命[136]。当前，中国已建立起门类齐全、层级清晰的生态环境法律体系，制定了生态环境保护相关法律30余部，初步建立了生态环境监测网络体系和生态环境监测评估技术体系，环境防治建设初见成效。实行环境保护50年，中国的环境保护目标已经从原来的总量控制，变成了以提高环境质量为重点。另外，以宪法为核心的环境保护法律制度得到了进一步完善，在世界范围内的引领性也得到了很大程度的提高，全国上下一条心，中国环境法治进入了新时代[136]。

改革开放后，中国用40年走完了发达国家用200年完成的工业化进程，虽然在如何处理经济发展与环境保护问题上有过迷茫困惑，有过思索反省，但也在不断调整修正着认知和实践。回顾中国的环保政策发展，可以看到环境管理体系实现了从单一到完备、从定性到定量、从抽象到具体的转变，具有明显的阶段性特征，许多环境保护法规实施的经验和教训为准确预测未来提供了警示。然而，现有的政策或方法还不足以应对中国未来面临的挑战。鉴于严峻的环境形势，中国政府应制定灵活的政策，同时开展广泛的国际合作，调动公众的参与。

1.3 环境保护难点与现有保护方法的局限性

中国已经将环境保护作为生态环境保护的重点工作，但与发达国家和地区相比，中国污染物风险防范和治理工作尚处于起步阶段，虽逐步取得了一些成效，但与有效防范污染物风险的目标要求仍存在较大差距，仍面临着客观上治理难度大、底数不清、能力不足的实际困难，也存在着治理体系不完善、权责不明、缺乏统筹协调、重末端治理轻源头治理和过程治理等问题。

1.3.1 环境保护难点

污染物治理之难，主要体现在五个方面：

（1）危害严重

各类有毒污染物会对人类的器官、神经、生殖发育等造成难以治愈的危害，给生态环境和人体健康带来了巨大风险。

（2）风险隐蔽

环境中大多数污染物在短期内对人类的毒害效应不明显，但是它们会通过食物链在人体内积累，从而对人的身体健康产生较大危害。

（3）不易处理

大量的污染物在环境中难以自然降解并易在生态系统中富集，通过食物链可长期在环境中和生物体内积累，常规的处理手段难以完全处理这些污染物。

（4）来源广泛

由于现代化工业发展迅速，随之而来的污染物来源极其广泛，可来源于人类在不经意间产生的污染物和环境降解产物。

（5）治理过程复杂

部分污染物难以被自然降解并且能够在生物体内富集，即使以很低的剂量排放到环境中，也可能会在长期积累的累积效应下危害环境、生物和人体健康，而对其进行完全处理往往涉及多个过程。

1.3.2 现有保护方法的局限性

众所周知，污染是人类在21世纪面临的全球性危机，大量证据表明，污染已导致人类出现各种健康问题[147-150]。随着工业的迅速发展和各类药品的大量生产使用，许多污染物对生态环境和公众健康的危害逐渐呈现在公众眼前。由于污染的范围非常广泛，污染治理往往达不到预期的目标。当前的物理、化学、生物联合技术均无法有效治理环境中多种多样的污染物，不经处理或有效处理会使得新污染物直接暴露在环境中，这将会导致更为严重的环境污染事件。有机污染物、重金属和其他新出现的污染物等会通过空气、水和食物进入每个人的日常生活，如果不采取强有力的措施来进行针对性处理，那么每个人的健康情况将会变得越来越糟。水体污染与土壤污染是污染治理的两大难点，现有的环境污染修复技术根据其修复机制的不同主要可分为物理修复、化学修复和生物修复，如表1-2和表1-3分别总结了当前土壤与水体污染治理措施的优势与不足。

表1-2　土壤污染修复技术的优势与不足

修复技术		措施	优点	缺点
物理修复	土壤置换	用未污染土壤替代污染土壤	适用于小面积、小体积重污染浅层土壤	成本高，污染土壤需要进一步处理
	固化	利用高温熔化土壤，并在冷却后固化污染物	可永久性治理，长期有效，应用范围广	成本高，功率损耗大，不适合大面积修复
	电动修复	在含有饱和污染土壤的电解槽两侧施加电流	修复时间短，能耗低，修复完全	处理深度有限，会降低土壤非均质物质的有效性
	热处理	通过蒸汽、微波和红外辐射加热受污染的土壤，使污染物挥发	工艺简单	投资成本高，需要控制气体排放，土壤结构容易被破坏

续表1-2

修复技术		措施	优点	缺点
化学修复	固化	在污染土壤中添加固定剂降低污染物在土壤中的流动性	成本低,简单,快速	不能去除土壤中的重金属,会改变土壤的理化性质
	稳定化	在污染土壤中添加试剂,将有毒废物转化为更稳定的形式	成本低,使用方便,抗生物降解性强,工程适用性好	需长期监测
	土壤洗涤	用各种试剂和萃取剂从土壤基质中浸出污染物	永久去除土壤中的污染物,可实现快速、高效地清理污染土壤	导致土壤结构恶化,修复成本高
生物修复	植物修复	使用根部吸附、沉淀和根区络合等特定机制降低污染物的流动性	性价比高,无创,无二次污染	修复能力和处理深度有限,修复周期长,植物和土壤需要长期监测
	微生物修复	利用微生物(如细菌、真菌和藻类)诱导土壤中污染物的吸附、沉淀、氧化和还原	成本低,修复能力强,对土壤结构、生态环境影响小,无二次污染	稳定性差,只能去除部分污染物

表1-3 水体污染修复技术的优势与不足

修复技术	措施	优点	缺点
物理修复	通过物理沉淀、过滤、吸附等作用除去可沉淀的固体、胶体、BODs、氮、磷、重金属、细菌、病毒及难以溶解的有机物质	操作简单,成本低	只能治理部分污染物,且治理不彻底
化学修复	通过向水体添加化学药剂和吸附剂改变水体氧化还原电位、pH,从而达到吸附沉淀水体中悬浮物质和有机质的目的;利用污染物的化学反应来使得污水中的有害物质转化为无害物质	操作简单,见效快	需大量添加化学药剂,成本高,同时容易引起二次污染

续表1-3

修复技术	措施	优点	缺点
生物修复	利用各种生物(包括微生物、动物和植物)的生理特性处理、吸收、降解、转化环境中的污染物,使受污染的环境得到改善	成本低,环境影响小,处理范围广	对水体修复的宏观效果不佳,稳定性差,对污染物的降解程度有限

环境污染治理是一个长期而又漫长的过程，在该过程中不可避免地又会有污染物随着人类活动排放到自然环境中，这些都将给公众健康带来极大的风险。大多数污染物可通过饮食摄入体内，导致污染物在人体积累，从而使得人体健康受到威胁，因此人们必须关注环境污染给公众健康带来的影响。

1.4 新时代中国环境治理方向

环境治理是全球发展的一个重大议题，对于进入新时代的中国尤为重要。近期召开的党的二十大，以习近平同志为核心的党中央总结过去，展望未来，进一步阐述了生态文明建设在建设中国式现代化中的重要性，中国环境保护事业迎来了发展的新思路与新任务。中国环境治理起步是比较晚的，周恩来等一众高瞻远瞩的中国领导人吸取发达国家经验教训，在环境问题还不太严重的情况下，开始抓环境保护，避免了重走发达国家“先污染后治理”的老路[131]。经过近50年的努力，中国在环境污染防治和自然生态保护方面取得了令人瞩目的成就，蓝天碧水已成为常态，美丽中国蓝图徐徐展开。当今的中国正在经历从一个环境保护的被动接受者到主动参与者再向重要引领者的转变。展望未来，我们应当按照党中央的新要求，重新谋划新时代的环境治理保护工作，奋力推进环境治理迈上新台阶。

1.4.1 完善环境监测指标

环境风险评估向来具有不确定性，这使得环境治理发展受到了一定限制。环境风险评估效果取决于输入数据的质量，因此科学的数据采集对于环境风险评估至关重要。环境监测是世界监测的重要组成部分，如空气质量、水质和土壤质量监测，也有一些新修订的法律法规加强了对污染物排放的监测，这些环保法规也为中国未来的污染改善提供了详细的指导。以重金属污染为例，重金属污染的人为来源包括采矿、冶炼、化石燃料燃烧、废物处理、腐蚀和农业生产等。随着工业的发展，中国也在遭受重金属污染，工业废水灌溉导致重金属污染了大片耕地，每年粮食污染量达数千万吨，人民的生命健康受到了威胁、财产受到了损害[151]。因此，进行土壤污染物的实时监测并及时预防调整是十分重要的。中国有关于土壤污染物的监测标准，但存在更新不及时、标准不统一、监测方法老旧等问题[151]。在环境治理工作中，环境监测是一个非常重要的部分。在目前尚不健全的监测系统面前，我们必须要不断地对监测标准进行更新，并加大对监测人员的训练和监管力度，同时还要重视对监测手段的发展和创新，这样，中国环境治理才能迈向更高的水平。

1.4.2 完善法律制度

从《中华人民共和国环境保护法（试行）》开始，中国的环境立法经过40多年的发展，从无到有，已全面覆盖各种环境和自然资源，以及与勘探、开发、开采、加工、利用和处置相关的活动，为中国的环境治理工作贡献了突出的力量。然而，在某些地区，环境治理仍然很棘手，需要继续完善和制定更加具体的环境保护政策法规。随着中国新一届领导人提出生态文明建设和新的国家发展战略，中国的环境立法将进入一个全新的发展阶段[120]。回顾过去环境治理取得的成就，我们可以断定中国未来可持续发展离不开法律保障。而目前中国环境立法仍存在许多缺陷和不足，如存在诸多空白，法律法规不协调现象突出，法律内容缺乏可操作性等[120]。展望未来，我们要建立政策评价技术体系，注重目标分析，加强前瞻性和预测性分析，完善新时代生态文明体系，建立科学高效的生态环境政策体系。

1.4.3 鼓励公众参与

近年来，随着公众环保意识的不断提高，无论是环保政策的制定中还是环保政策的实施中都开始出现公众的声音，公众参与的力量越来越无法被忽视。在中国，公众参与环境立法尤为重要，因为目前的立法实体需要更多的技术援助来规范环境政策，公众利益在立法机构中的代表性相对较弱，公众参与环境保护法的实施需要更好的制度支持。中国通过部分法律法规，已经实现了公众参与环境决策一定程度的制度化。比如，2007年，国家环保总局（现环境保护部）颁布了《环境信息公开办法（试行）》，要求各级环保部门公开政府环境信息，而且强制污染严重的企业公开有关信息，在保障公民最基本的知情权的同时促进了公众参与污染治理监督。但这些政策的实际效果并不令人满意，参与方式单一、获取信息困难等问题仍然存在，阻碍了有意义的公众参与。发展公众参与环境治理工作，需要我们未来通过更好地利用现有制度和逐步完善机制来改善参与制度缺失的问题，同时还要让公众更明白参与环境治理的意义，拓宽公众参与的渠道，呼吁更多人参与其中。另外，新时代互联网公众参与已成为监管企业污染活动的有力工具[152]，政府要重视官方微博账号建设，关注公众提出的问题，并及时反馈。

1.4.4 加强国际合作

保护地球环境是一个全球性事业，开展广泛的国际合作是其必然趋势。环境保护不应受国界限制，全世界需要团结起来，积极应对环境污染，保护生态环境。为此，必须展开深度的国际合作：一方面可以利用国际科学研究和管理经验来发展中国环境治理工作；另一方面，作为联合国安理会五个常任理事国之一，中国应积极承担环境保护责任，树立大国形象。中国作为最大的发展中国家，应自觉、按期履行减排义务，承担环境治理的国际责任，这也是中国对世界的庄严承诺[143]。党的十九大，习近平总书记发出了中国要作“全球生态文明建设的重要参与者、贡献者、引领者”的号召，表明了中国承担环境治理的国际责任和决心。

（葛斌　邓杰芳　云慧*①）

参考文献

[1]杨化菊，李婷，李颂群，等. 环境污染对动物遗传多样性的影响研究进展[J]. 环境生态学，2021，3(1)：49-56.

[2]LIU Q, LONG Y, WANG C, et al. Drivers of provincial SO_2 emissions in China-based on multi-regional input-output analysis[J]. Journal of cleaner production, 2019, 238(20): 78-93.

[3]CLEUVERS M. Aquatic ecotoxicity of pharmaceuticals including the assessment of combination effects[J]. Toxicology letters, 2003, 142(3): 185-194.

[4]FERRARI B, PAXEUS N, LO GIUDICE R, et al. Ecotoxicological impact of pharmaceuticals found in treated wastewaters: study of carbamazepine, clofibric acid, and diclofenac[J]. Ecotoxicology and environmental safety, 2003, 55(3): 359-370.

[5]AL AUKIDY M, VERLICCHI P, JELIC A, et al. Monitoring release of pharmaceutical compounds: occurrence and environmental risk assessment of two WWTP effluents and their receiving bodies in the Po Valley, Italy[J]. Science of the total environment, 2012, 438: 15-25.

[6]VERLICCHI P, AL AUKIDY M, ZAMBELLO E. Occurrence of pharmaceutical compounds in urban wastewater: removal, mass load and environmental risk after a

① 带*的为通信作者，无*的为共同作者。

secondary treatment - a review[J]. Science of the total environment, 2012, 429: 123–155.

[7]VERLICCHI P, ZAMBELLO E. Pharmaceuticals and personal care products in untreated and treated sewage sludge: occurrence and environmental risk in the case of application on soil - a critical review[J]. Science of the total environment, 2015, 538: 750–767.

[8]CHEN R S, YE C. Resolving soil pollution in China[J]. Nature, 2014, 505(7484):483.

[9]ZHANG C, ZENG G M, HUANG D L, et al. Combined removal of di(2-ethylhexyl) phthalate (DEHP) and Pb(Ⅱ) by using a cutinase loaded nanoporous gold-polyethyleneimine adsorbent[J]. RSC Advances, 2014, 4(98): 55511–55518.

[10]ZHANG C, LAI C, ZENG G M, et al. Efficacy of carbonaceous nanocomposites for sorbing ionizable antibiotic sulfamethazine from aqueous solution[J]. Water research, 2016, 95: 103–112.

[11]JÄRUP L. Hazards of heavy metal contamination[J]. British medical bulletin, 2003, 68(1): 167–182.

[12]HE B, YUN Z, SHI J, et al. Research progress of heavy metal pollution in China: sources, analytical methods, status, and toxicity[J]. Chinese science bulletin, 2013, 58(2): 134–140.

[13]WELLING R, BEAUMONT J J, PETERSEN S J, et al. Chromium Ⅳ and stomach cancer: a meta-analysis of the current epidemiological evidence[J]. Occupational and environmental medicine, 2015, 72(2): 151–159.

[14]WU Y, PANG H, LIU Y, et al. Environmental remediation of heavy metal ions by novel-nanomaterials: a review[J]. Environmental pollution, 2019, 246: 608–620.

[15]PINTO M, MARINHO-REIS P, ALMEIDA A, et al. Links between cognitive status and trace element levels in hair for an environmentally exposed population: a case study in the surroundings of the estarreja industrial area[J]. International journal of environmental research and public health, 2019, 16(22): 45–60.

[16]PRASAD S, YADAV K K, KUMAR S, et al. Chromium contamination and effect on environmental health and its remediation: a sustainable approaches[J]. Journal

of environmental management, 2021, 285: 1-22.

[17] ZHOU Q Q, YANG N, LI Y Z, et al. Total concentrations and sources of heavy metal pollution in global river and lake water bodies from 1972 to 2017[J]. Global ecology and conservation, 2020, 22: 1-11.

[18] VARDHAN K H, KUMAR P S, PANDA R C. A review on heavy metal pollution, toxicity and remedial measures: current trends and future perspectives[J]. Journal of molecular liquids, 2019, 290: 1-22.

[19] LE T T N, LE V T, DAO M U, et al. Preparation of magnetic graphene oxide/chitosan composite beads for effective removal of heavy metals and dyes from aqueous solutions[J]. Chemical engineering communications, 2019, 206(10): 1337-1352.

[20] HU B, SHAO S, NI H, et al. Current status, spatial features, health risks, and potential driving factors of soil heavy metal pollution in China at province level[J]. Environmental pollution, 2020, 266(3): 1-21.

[21] BERNHOFT R A. Cadmium toxicity and treatment[J]. Scientific world journal, 2013, 2013: 1-7.

[22] ZOU M, ZHOU S, ZHOU Y, et al. Cadmium pollution of soil-rice ecosystems in rice cultivation dominated regions in China: a review[J]. Environmental pollution, 2021, 280: 1-15.

[23] SHI J J, SHI Y, FENG Y L, et al. Anthropogenic cadmium cycles and emissions in mainland China 1990-2015[J]. Journal of cleaner production, 2019, 230: 1256-1265.

[24] BERNARD A. Cadmium & its adverse effects on human health[J]. Indian journal of medical research, 2008, 128(4): 557-564.

[25] JIN Y, LIU Z, LIU F, et al. Embryonic exposure to cadmium (Ⅱ) and chromium (Ⅵ) induce behavioral alterations, oxidative stress and immunotoxicity in zebrafish (*Danio rerio*)[J]. Neurotoxicology and teratology, 2015, 48: 9-17.

[26] ALGHASHAM A, SALEM T A, MEKI A R M. Effect of cadmium-polluted water on plasma levels of tumor necrosis factor-alpha, interleukin-6 and oxidative status biomarkers in rats: protective effect of curcumin[J]. Food and chemical toxicology, 2013, 59: 160-164.

[27] JIN Y X, LIU L, ZHANG S B, et al. Cadmium exposure to murine macro-

phages decreases their inflammatory responses and increases their oxidative stress[J]. Chemosphere,2016,144(12):168–175.

[28]KE S,KE Q M,JIA W J,et al. Benchmark dose estimation for cadmium-induced renal effects based on a large sample population from five Chinese provinces [J]. Biomedical and environmental sciences,2015,28(5):383–387.

[29]CHEN Q Y,DESMARAIS T,COSTA M. Metals and mechanisms of carcinogenesis[J]. Annu Rev Pharmacol Toxicol,2019,59(1):537–554.

[30]ZEWDU F, AMARE M. Determination of the level of hexavalent, trivalent, and total chromium in the discharged effluent of bahir dar tannery using ICP-OES and UV - Visible spectrometry[J]. Cogent chemistry,2018,4(1):1–9.

[31]YOUNAN S,SAKITA G Z,ALBUQUERQUE T R,et al. Chromium(Ⅵ)bioremediation by probiotics[J]. Journal of the science of food and agriculture, 2016, 96 (12):3977–3982.

[32]JAISHANKAR M,TSETEN T,ANBALAGAN N,et al. Toxicity, mechanism and health effects of some heavy metals[J]. Interdisciplinary toxicology, 2014, 7(6): 60–72.

[33]MURTHY M K,KHANDAYATARAY P,PADHIARY S,et al. A review on chromium health hazards and molecular mechanism of chromium bioremediation[J]. Reviews on environmental health,2023,38(3):461–478.

[34]OZUAH P O. Mercury poisoning[J]. Current problems in pediatrics,2000,30 (4):91–99.

[35]DURUIBE J O,OGWUEGBU M O C,EGWURUGWU J N. Heavy metal pollution and human biotoxic effects[J]. International journal of the physical sciences, 2007,2(5):112–118.

[36]TAKIZAWA Y. Understanding minamata disease and strategies to prevent further environmental contamination by methylmercury[J]. Water science and technology,2000,42(7–8):139–146.

[37]HUI C,GUO Y,ZHANG W,et al. Surface display of PbrR on Escherichia coli and evaluation of the bioavailability of lead associated with engineered cells in mice [J]. Scientific reports,2018,8(1):56–85.

[38]VAN DER KUIJP T J,HUANG L,CHERRY C R. Health hazards of China's

lead - acid battery industry: a review of its market drivers, production processes, and health impacts[J]. Environmental health, 2013, 12(1/2): 1–10.

[39]TAYLOR M P, FORBES M K, OPESKIN B, et al. The relationship between atmospheric lead emissions and aggressive crime: an ecological study[J]. Environmental health, 2016, 15(22): 1–10 .

[40]HALLAJI H, KESHTKAR A R, MOOSAVIAN M A. A novel electrospun PVA/ZnO nanofiber adsorbent for U(Ⅵ), Cu(Ⅱ) and Ni(Ⅱ) removal from aqueous solution[J]. Journal of the Taiwan institute of chemical engineers, 2015, 46: 109–118.

[41]ATARI L, ESMAEILI S, ZAHEDI A, et al. Removal of heavy metals by conventional water treatment plants using poly aluminum chloride[J]. Toxin reviews, 2019, 38(2): 127–134.

[42]ZHUANG P, MCBRIDE M B, XIA H P, et al. Health risk from heavy metals via consumption of food crops in the vicinity of Dabaoshan mine, South China[J]. Science of the total environment, 2009, 407(5): 1551–1561.

[43]ROGIVAL D, SCHEIRS J, BLUST R. Transfer and accumulation of metals in a soil-diet-wood mouse food chain along a metal pollution gradient[J]. Environmental pollution, 2007, 145(2): 516–528.

[44]GARCIA F E, SENN A M, MEICHTRY J M, et al. Iron-based nanoparticles prepared from yerba mate extract. Synthesis, characterization and use on chromium removal[J]. Journal of environmental management, 2019, 235: 1–8.

[45]LEONG Y K, CHANG J S. Bioremediation of heavy metals using microalgae: recent advances and mechanisms[J]. Bioresource technology, 2020, 303: 1–11.

[46]胡延春，贾艳，张乃生. 生物毒素的应用研究[J]. 生物技术通讯，2004(1): 83–85.

[47]王龑，王刘庆，刘阳. 食品中主要真菌毒素生物合成途径研究进展[J]. 食品安全质量检测学报，2016, 7(6): 2158–2167.

[48]KHLANGWISET P, SHEPHARD G S, WU F. Aflatoxins and growth impairment: a review[J]. Critical reviews in toxicology, 2011, 41(9): 740–755.

[49]PINTON P, OSWALD I P. Effect of deoxynivalenol and other type B trichothecenes on the intestine: a review [J]. Toxins, 2014, 6(5): 1615–1643.

[50]PAYROS D, ALASSANE K I, PIERRON A, et al. Toxicology of deoxynivalenol and its acetylated and modified forms[J]. Archives of toxicology, 2016, 90(12): 2931–2957.

[51]MARESCA M, FANTINI J. Some food - associated mycotoxins as potential risk factors in humans predisposed to chronic intestinal inflammatory diseases[J]. Toxicon, 2010, 56(3): 282–294.

[52]KAKADE A, SALAMA E S, HAN H, et al. World eutrophic pollution of lake and river: biotreatment potential and future perspectives[J]. Environmental technology & innovation, 2021, 23: 1–23.

[53]HITZFELD B C, HOGER S J, DIETRICH D R. Cyanobacterial toxins: removal during drinking water treatment, and human risk assessment [J]. Environmental health perspectives, 2000, 108(1): 113–122.

[54]RAO P V L, GUPTA N, BHASKAR A S B, et al. Toxins and bioactive compounds from cyanobacteria and their implications on human health[J]. Journal of environmental biology, 2002, 23(3): 215–224.

[55]刘沛,黄慧敏,余涛,等.我国新污染物污染现状、问题及治理对策[J].环境监控与预警,2022,14(5):27–30.

[56]SIDHU J P S, AHMED W, GERNJAK W, et al. Sewage pollution in urban stormwater runoff as evident from the widespread presence of multiple microbial and chemical source tracking markers [J]. Science of the total environment, 2013, 463: 488–496.

[57]JONES O A H, VOULVOULIS N, LESTER J N. Potential ecological and human health risks associated with the presence of pharmaceutically active compounds in the aquatic environment[J]. Critical reviews in toxicology, 2004, 34(4): 335–350.

[58]MICHAEL I, RIZZO L, MCARDELL C S, et al. Urban wastewater treatment plants as hotspots for the release of antibiotics in the environment: a review[J]. Water research, 2013, 47(3): 957–995.

[59]LUO Y, GUO W, NGO H H, et al. A review on the occurrence of micropollutants in the aquatic environment and their fate and removal during wastewater treatment [J]. Science of the total environment, 2014, 473: 619–641.

[60]KAUR R, MAVI G, RAGHAV S, et al. Pesticides classification and its im-

pact on environment[J]. International journal of current microbiology and applied sciences,2019,8:1889-1897.

[61]KARUNARATHNE A,GUNNELL D,KONRADSEN F,et al. How many premature deaths from pesticide suicide have occurred since the agricultural green revolution?[J]. Clinical toxicology,2020,58(8):227-232.

[62]THAKUR D S,KHOT R,JOSHI P P,et al. Glyphosate poisoning with acute pulmonary edema[J]. Toxicology international,2014,21(3):328-330.

[63]MNIF W,HASSINE A I H,BOUAZIZ A,et al. Effect of endocrine disruptor pesticides: a review [J]. International journal of environmental research and public health,2011,8(6):2265-2303.

[64]KIM K H,KABIR E,JAHAN S A. Exposure to pesticides and the associated human health effects[J]. Science of the total environment,2017,575:525-535.

[65]EL G K S,ALY N M,MAHMOUD F H,et al. The role of vitamin C as antioxidant in protection of oxidative stress induced by imidacloprid[J]. Food and chemical toxicology,2010,48(1):215-221.

[66]SOHN H Y,KUM E J,KIM J S,et al. Toxicity evaluation of organochloride pesticide,endosulfan and its metabolites using microalgae[J]. Korean journal of microbiology and biotechnology,2006,34(4):357-362.

[67]CHHILLAR N,SINGH N K,BANERJEE B D,et al. Organochlorine pesticide levels and risk of Parkinson's disease in north Indian population[J]. ISRN neurology,2013,2013:1-6.

[68]BALALI M M,SHARIAT M. Treatment of organophosphate poisoning. Experience of nerve agents and acute pesticide poisoning on the effects of oximes[J]. Journal of physiology-paris,1998,92(5):375-378.

[69]GUO J,LUO Y,LI H,et al. Sensitive fluorescent detection of carbamate pesticides represented by methomyl based on the inner filter effect of Au nanoparticles on the fluorescence of CdTe quantum dots[J]. Analytical methods,2013,5(23):6830-6838.

[70]KARAMI M S,ABDOLLAHI M. Toxic influence of organophosphate,carbamate,and organochlorine pesticides on cellular metabolism of lipids,proteins,and carbohydrates: a systematic review[J]. Human & experimental toxicology,2011,30(9):

1119–1140.

[71]YAN X, KONG D, JIN R, et al. Fluorometric and colorimetric analysis of carbamate pesticide via enzyme-triggered decomposition of gold nanoclusters-anchored MnO_2 nanocomposite[J]. Sensors and actuators B: chemical, 2019, 290: 640–647.

[72]WANG X, HOU T, DONG S, et al. Fluorescence biosensing strategy based on mercury ion-mediated DNA conformational switch and nicking enzyme-assisted cycling amplification for highly sensitive detection of carbamate pesticide[J]. Biosensors and bioelectronics, 2016, 77: 644–649.

[73]YANG N, WANG P, XUE C Y, et al. A portable detection method for organophosphorus and carbamates pesticide residues based on multilayer paper chip[J]. Journal of food process engineering, 2018, 41(8): 1–10.

[74]CHRUSTEK A, HOLYNSKA-IWAN I, DZIEMBOWSKA I, et al. Current research on the safety of pyrethroids used as insecticides [J]. Medicina-lithuania, 2018, 54(4): 1–15.

[75]ORSBORNE J, BANKS S D, HENDY A, et al. Personal protection of permethrin-treated clothing against *Aedes aegypti*, the vector of dengue and Zika virus, in the laboratory [J]. PLoS One, 2016, 11(5): 1–18.

[76]RANJKESH M R, NAGHILI B, GOLDUST M, et al. The efficacy of permethrin 5% vs oral ivermectin for the treatment of scabies [J]. Annals of parasitology, 2013, 59(4): 189–194.

[77]CORCELLAS C, LUISA FEO M, PAULO TORRES J, et al. Pyrethroids in human breast milk: occurrence and nursing daily intake estimation[J]. Environment international, 2012, 47: 17–22.

[78]WHITEHOUSE C R, BOULLATA J, MCCAULEY L A. The potential toxicity of artificial sweeteners [J]. Aaohn journal, 2008, 56(6): 251–259.

[79]GAN Z, SUN H, FENG B, et al. Occurrence of seven artificial sweeteners in the aquatic environment and precipitation of Tianjin, China[J]. Water research, 2013, 47(14): 4928–4937.

[80]AMY SAGERS C, REINHARDT K, LARSON D M. Ecotoxicological assessments show sucralose and fluoxetine affect the aquatic plant, *Lemna minor*[J]. Aquatic toxicology, 2017, 185: 76–85.

[81]AZAD M B, ABOU S A M, CHAUHAN B F, et al. Nonnutritive sweeteners and cardiometabolic health: a systematic review and meta-analysis of randomized controlled trials and prospective cohort studies[J]. Canadian medical association journal, 2017, 189(28): 929-939.

[82]RENWICK A G. The intake of intense sweeteners-an update review[J]. Food additives and contaminants part a-chemistry analysis control exposure & risk assessment, 2006, 23(4): 327-338.

[83]门辛. 甜味剂的分类[J]. 标准计量与质量, 2001(S1): 39.

[84]SCHIFFMAN S S, ROTHER K I. Sucralose, a synthetic organochlorine sweetener: overview of biological issues [J]. Journal of toxicology and environmental health-part b-critical reviews, 2013, 16(7): 399-451.

[85]祝海珍. 乳制品中三聚氰胺检测方法的现状及研究进展[J]. 食品安全质量检测学报, 2021, 12(3): 1009-1014.

[86]MAST R W, JEFFCOAT A R, SADLER B M, et al. Investigating the interaction between melamine and cyanuric acid using a physiologically based toxicokinetic model in rainbow trout [J]. Food and chemical toxicology, 1983, 21: 807-810.

[87]MELNICK R L, BOORMAN G A, HASEMAN J K, et al. Urolithiasis and bladder carcinogenicity of melamine in rodents [J]. Toxicology and applied pharmacology, 1984, 72(2): 292-303.

[88]OGASAWARA H, IMAIDA K, ISHIWATA H, et al. Urinary bladder carcinogenesis induced by melamine in F344 male rats: correlation between carcinogenicity and urolith formation [J]. Carcinogenesis, 1995, 16(11): 2773-2777.

[89]ASCHEBROOK KILFOY B, CROSS A J, STOLZENBERG SOLOMON R Z, et al. Pancreatic cancer and exposure to dietary nitrate and nitrite in the NIH-AARP diet and health study[J]. American journal of epidemiology, 2011, 174(3): 305-315.

[90]KUMAR M, JAISWAL S, SODHI K K, et al. Antibiotics bioremediation: perspectives on its ecotoxicity and resistance[J]. Environment international, 2019, 124: 448-461.

[91]RONQUILLO M G, HERNANDEZ J C A. Antibiotic and synthetic growth promoters in animal diets: review of impact and analytical methods[J]. Food control, 2017,

72:255–267.

[92]COWIESON A J,KLUENTER A M. Contribution of exogenous enzymes to potentiate the removal of antibiotic growth promoters in poultry production[J]. Animal feed science and technology,2019,250:81–92.

[93]CYCON M,MROZIK A,PIOTROWSKA SEGET Z. Antibiotics in the soil environment - degradation and their impact on microbial activity and diversity[J]. Frontiers in microbiology,2019,10:1–45.

[94]CHEN Y,SU J Q,ZHANG J,et al. High-throughput profiling of antibiotic resistance gene dynamic in a drinking water river-reservoir system[J]. Water research,2019,149:179–189.

[95]ZHU Y G,JOHNSON T A,SU J Q,et al. Diverse and abundant antibiotic resistance genes in Chinese swine farms[J]. Proceedings of the national academy of sciences,2013,110(9):3435–3440.

[96]LORENZO P,ADRIANA A,JESSICA S,et al. Antibiotic resistance in urban and hospital wastewaters and their impact on a receiving freshwater ecosystem[J]. Chemosphere,2018,206:70–82.

[97]SHAO S,HU Y,CHENG J,et al. Research progress on distribution,migration,transformation of antibiotics and antibiotic resistance genes (ARGs)in aquatic environment[J]. Critical reviews in biotechnology,2018,38(8):1195–1208.

[98]LIU X H,LU S Y,GUO W,et al. Antibiotics in the aquatic environments:a review of lakes,China[J]. Science of the total environment,2018,627:1195–1208.

[99]DANNER M C,ROBERTSON A,BEHRENDS V,et al. Antibiotic pollution in surface fresh waters:occurrence and effects[J]. Science of the total environment,2019,664:793–804.

[100]KOVALAKOVA P,CIZMAS L,MCDONALD T J,et al. Occurrence and toxicity of antibiotics in the aquatic environment:a review[J]. Chemosphere,2020,251:1–15.

[101]KELSIC E D,ZHAO J,VETSIGIAN K,et al. Counteraction of antibiotic production and degradation stabilizes microbial communities[J]. Nature,2015,521(7553):516–519.

[102]PRUDEN A,PEI R T,STORTEBOOM H,et al. Antibiotic resistance genes

as emerging contaminants: studies in northern Colorado[J]. Environmental science & technology,2006,40(23):7445-7450.

[103]ZHANG Y,CHENG D M,XIE J,et al. Impacts of farmland application of antibiotic-contaminated manures on the occurrence of antibiotic residues and antibiotic resistance genes in soil:a meta-analysis study[J]. Chemosphere,2022,300:1-18.

[104]BIELEN A,SIMATOVIC A,KOSIC VUKSIC J,et al. Negative environmental impacts of antibiotic-contaminated effluents from pharmaceutical industries[J]. Water research,2017,126:79-87.

[105]BINH V N,DANG N,ANH N T K,et al. Antibiotics in the aquatic environment of Vietnam: sources, concentrations, risk and control strategy[J]. Chemosphere,2018,197:438-450.

[106]梅筱玲.克拉霉素的副作用及护理对策[J].现代医药卫生,2001(9):753.

[107]蒋克春.甲硝唑的毒副作用及应用注意事项[J].航空军医,2003(4):183-184.

[108]陈伟.环丙沙星的不良反应[J].药物与人,2002(7):47.

[109]凌晓敏.对克林霉素的副作用及用药安全性的分析[J].当代医药论丛,2015,13(13):149-150.

[110]曹文华,崔怀西.β-内酰胺类抗生素的毒副作用[J].江苏医药,1994(11):619-620.

[111]LIN S,LI Y,LI Y H,et al. Influence of tree size,local forest structure,topography, and soil resource availability on plantation growth in Qinghai Province, China[J]. Ecological Indicators,2021,120(C):1-8.

[112]SONG B,ZENG G M,GONG J L,et al. Evaluation methods for assessing effectiveness of in situ remediation of soil and sediment contaminated with organic pollutants and heavy metals[J]. Environment international,2017,105:43-55.

[113]FENT K,WESTON A A,CAMINADA D. Ecotoxicology of human pharmaceuticals [J]. Aquatic toxicology,2006,78(2):122-159.

[114]CIOCCO A,THOMPSON D J. A follow-up of donora ten years after:methodology and findings[J]. American journal of public health and the nation's health,1961,51(2):155-164.

[115]GLENCROSS D A,HO T R,CAMILLA N,et al. Air pollution and its effects on the immune system[J]. Free radical biology and medicine,2020,151:56–68.

[116]佚名.全球十大环境污染事件[J].环境教育,2004(2):32.

[117]佚名.日本米糠油事件——生产污染危及生命安全[J].现代班组,2022(4):28.

[118]佚名.印度博帕尔毒气泄漏事件[J].世界环境,2012(1):7.

[119]佚名.莱茵河污染事故[J].世界环境,2009(6):7.

[120]MU Z, BU S, XUE B. Environmental legislation in China: achievements, challenges and trends[J]. Sustainability,2014,18(6):8967–8979.

[121]付战勇,马一丁,罗明,等.生态保护与修复理论和技术国外研究进展[J].生态学报,2019,39(23):9008–9021.

[122]元淼.浅析环境工程管理中存在的问题与对策[J].科技风,2021(24):113–115.

[123]冯玮.试论政府环境责任[J].知识经济,2013(8):10.

[124]SUN X,PING Z B,DONG Z F,et al. Resources and environmental costs of China's rapid economic growth:from the latest theoretic SEEA framework to modeling practice[J]. Journal of cleaner production,2021,315:1–17.

[125]LU Y L,ZHANG Y Q,CAO X H,et al. Forty years of reform and opening up:China's progress toward a sustainable path[J]. Science advances,2019,5(8):1–10.

[126]XIE R,ZHAO G,ZHU B Z,et al. Examining the factors affecting air pollution emission growth in China[J]. Environmental modeling & assessment,2018,23(4):389–400.

[127]HUANG Y,MI F,WANG J,et al. Water pollution incidents and their influencing factors in China during the past 20 years[J]. Environmental monitoring and assessment,2022,194(3):17–29.

[128]郑少华,王慧.中国环境法治四十年:法律文本、法律实施与未来走向[J].法学,2018(11):17–29.

[129]王金南,董战峰,蒋洪强,等.中国环境保护战略政策70年历史变迁与改革方向[J].环境科学研究,2019,32(10):1636–1644.

[130]王晓红,张亦工.中国环境政策体系:构建与发展——基于生态文明的

视角[J]. 山东财经大学学报,2021,33(5):43-50.

[131]曲格平. 九十感怀[J]. 环境经济,2020(17):10-13.

[132]罗理恒,张希栋,曹超. 中国环境政策40年历史演进及启示[J]. 环境保护科学,2022,48(4):34-38.

[133]汪自书,胡迪. 我国环境管理新进展及环境大数据技术应用展望[J]. 中国环境管理,2018,10(5):90-96.

[134]曲格平. 但见时光流似箭,砥砺前行书新篇——中国生态环境历程回顾与思考[J]. 中华环境,2019(10):22-26.

[135]孙佑海. 我国70年环境立法:回顾、反思与展望[J]. 中国环境管理,2019,11(6):5-10.

[136]吕忠梅,吴一冉. 中国环境法治七十年:从历史走向未来[J]. 中国法律评论,2019(5):102-123.

[137]ZHANG K M, WEN Z G. Review and challenges of policies of environmental protection and sustainable development in China[J]. Journal of environmental management, 2008, 88(4):1249-1261.

[138]吴荻,武春友. 建国以来中国环境政策的演进分析[J]. 大连理工大学学报(社会科学版),2006(4):48-52.

[139]马小明,赵月炜. 环境管制政策的局限性与变革——自愿性环境政策的兴起[J]. 中国人口·资源与环境,2005(6):19-23.

[140]段新,戴胜利,乔杰. 新中国成立以来环境管理体制的变迁:历程与逻辑[J]. 四川行政学院学报,2020(4):49-60.

[141]ROTH D. China's environomic challenge[J]. The journal of environment & development, 1994, 3(2):139-145.

[142]BRADBURY I, KIRKBY R. China's agenda 21: a critique[J]. Applied geography, 1996, 16(2):97-107.

[143]XU K, TIAN G. Codification and prospect of China & rsquo;s codification of environmental law from the perspective of global environmental governance[J]. International journal of environmental research and public health, 2022, 19(16):1-14.

[144]WANG L. The changes of China's environmental policies in the latest 30 years[J]. Procedia environmental sciences, 2010, 2:1206-1212.

[145]YANG Y, YAO C, XU D. Ecological compensation standards of national sce-

nic spots in western China: a case study of Taibai Mountain[J]. Tourism management, 2020, 76: 1-17.

[146]张小筠,刘戒骄.新中国70年环境规制政策变迁与取向观察[J].改革, 2019(10):16-25.

[147]CARPENTER D O. Health effects of persistent organic pollutants: the challenge for the Pacific Basin and for the world[J]. Reviews on environmental health 2011, 26(11): 61-69.

[148]NARAYANA K. An aminoglycoside antibiotic gentamycin induces oxidative stress, reduces antioxidant reserve and impairs spermatogenesis in rats[J]. The journal of toxicological sciences, 2008, 33(1): 85-96.

[149]POET T S, WU H, KOUSBA A A, et al. In vitro rat hepatic and intestinal metabolism of the organophosphate pesticides chlorpyrifos and diazinon[J]. Toxicological sciences, 2003, 72(2): 193-200.

[150]CHEN J, YING G G, DENG W J. Antibiotic residues in food: extraction, analysis, and human health concerns[J]. Journal of agricultural and food chemistry, 2019, 67(27): 7569-7586.

[151]汪志红.我国土壤环境监测的问题与对策研究[J].清洗世界, 2022, 38(6): 167-169.

[152]WU W, WANG W, ZHANG M. Does internet public participation slow down environmental pollution?[J]. Environmental science & policy, 2022, 137: 22-31.

第2章
污染物与肠道菌群的相互作用

2017年1月18日，国家主席习近平在联合国日内瓦总部做题为《共同构建人类命运共同体》的主旨演讲时提出："工业化创造了前所未有的物质财富，也产生了难以弥补的生态创伤。我们不能吃祖宗饭、断子孙路，用破坏性方式搞发展。绿水青山就是金山银山。我们应该遵循天人合一、道法自然的理念，寻求永续发展之路。"

2.1 污染物与肠道菌群

肠道是人及动物体内重要的消化器官，很多栖息在肠道里的微生物参与宿主的代谢及生理功能，这些微生物也因此被称为"肠道菌群"。肠道菌群由500～1000种微生物组成，其中大部分为厌氧菌，分布在肠道的各个部位，如图2-1所示，在结肠中微生物含量可达10^{12} CFU/mL[1]。肠道菌群在门水平上主要是厚壁菌门（Firmicutes）及拟杆菌门（Bacteroidetes），其次是放线菌门（Actinobacteria）和疣微菌门（Verrucomicrobia）[2]。厚壁菌门和拟杆菌门在健康的肠道菌群组成中扮演着重要的角色，其中厚壁菌门和拟杆菌门的比例与部分疾病相关[3]。在肥胖人群中厚壁菌和拟杆菌的比例要高于正常人的比例，这是由于厚壁菌门中一些微生物可以有效促进能量吸收，从而导致肥胖[4]。在结肠中双歧杆菌（*Bifidobacterium*）、乳酸杆菌（*Lactobacillus*）、拟杆菌（*Bacteroides*）、肠杆菌

(*Enterobacter*)等为主要优势菌，但是乳酸菌、嗜黏蛋白阿克曼氏菌(*Akkermansia muciniphila*)等部分菌属栖息在黏液层上，另外沙门氏菌(*Salmonella*)、脆弱拟杆菌(*Bacteroides fragilis*)等丰度较低的属基本为肠道致病菌[5]。这些栖息在肠道中的大部分菌属为肠道共生菌，其与宿主相互作用维持肠道系统的生态平衡。肠道共生菌的主要功能是对营养物质进行消化吸收，从而为宿主合成ATP提供能量[6]。而这些共生菌在肠道系统中也可以通过分泌黏蛋白从而形成强大的肠道黏膜屏障，进而抵抗外源致病菌的入侵[7, 8]。在结肠内含有的未消化的食物残渣，如膳食纤维，会在肠道菌群的作用下进行发酵，从而产生短链脂肪酸(short-chain fatty acids，SCFA)，如甲酸、乙酸、丁酸、丙酸等。这些短链脂肪酸为肠道共生菌的代谢物，可与宿主相互作用，参与宿主的免疫调节，

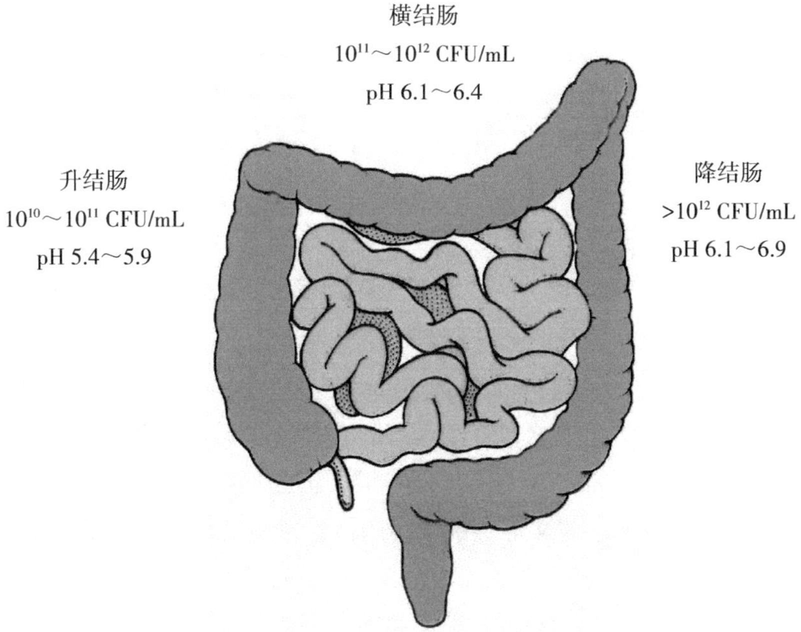

图2-1　人体肠道结构图

如刺激T细胞产生免疫蛋白及分泌抗炎因子[9]。另外，部分短链脂肪酸如丁酸与癌症的预防有关[10]。部分肠道共生菌可参与宿主体内的免疫调节，免疫细胞产生的免疫球蛋白A（immunoglobulin A，IgA）抗体不仅可以参与免疫反应，而且还能促进肠道共生菌细胞膜的生长[8]。在肠道共生菌中部分菌群可黏附在肠道黏液层上生长，其以肠道黏液层中的黏蛋白为碳源和氮源，从而减少了肠道上皮细胞的渗透性，并同时参与合成了黏液层多糖[11，12]。

肠道菌群在宿主内处于一个动态变化的过程之中，饮食、药物等外界因素会使其组成发生改变，另外各种疾病的发生都会导致肠道菌群结构组成的改变甚至导致稳态失衡。炎症性肠病（inflammatory bowel disease，IBD）患者出现肠道菌群物种多样性和稳定性降低的现象时，其肠道中厚壁菌门减少，拟杆菌门增加，同时部分兼性厌氧菌如肠杆菌相对丰度也出现了增加的现象[13，14]。脊髓损伤（spinal cord injury，SCI）患者粪便样本16S rDNA测序发现罗氏菌属（*Roseburia*）和假丁酸弧菌属（*Pseudobutyrivibrio*）等产丁酸菌的相对丰度降低[15，16]。在肥胖人群中，拟杆菌门中的一种多形类杆菌（*Bacteroides thetaiotaomicron*）的相对丰度显著降低，且与血清谷氨酸浓度呈负相关[17，18]。2型糖尿病患者肠道菌群中丁酸盐产生菌相对丰度下降，嗜黏蛋白阿克曼氏菌丰度下降，部分具有促炎潜力的菌丰度增加[18，19]。在妊娠糖尿病（gestational diabetes mellitus，GDM）患者中除了丁酸盐产生菌减少之外，肠道菌群丰度降低的同时致病菌丰度增加[18，20]。另外肠道菌群通过代谢产物，如胆汁酸、短链脂肪酸等直接或者间接性影响高胆固醇血症和冠状动脉疾病（coronary artery disease，CAD）的发生[21]。

随着工业的发展，环境污染物的存在越来越普遍，如重金属、室内甲醛、残留农药、食品微塑料及医用抗生素等。这些污染物不论是生活产生的，还是工业生产产生的，都对人体健康带来了致命的威胁。正常情况下，生活中的重金属含量都具有一定的安全范围，当其含量超标或在生物体内过多富集就会对生物体产生毒性作用。例如，六价铬含量超标会造成肝脏损伤及内部器官出血，同时会使人体的呼吸系统受到损伤[22]。人类在日常生活中接触最多的重金属污染是食源性污染，如重金属汞的暴露一般源自于鱼类及牙齿汞合金，汞的暴露首先会造成大脑受损，另外神经系统、免疫系统及肾脏等都会不同程度地受到损伤[23]。除了重金属污染之外，日常生活中有机物污染也是主要的污染物来源，如瓜果蔬菜中的农药残留及有机农药不当排放造成的饮用水污染，大量有机磷农药会导致人出现肺水肿并短期呼吸衰竭而死亡[24]。食源性污染物中微塑料污染也是常见的污染物来源，日常生活中所用的塑料制品不可避免地会使人体摄入微塑料，而微塑料会对人体消化系统、免疫系统、呼吸系统等产生毒性作用[25]。另外，抗生素的滥用会促使产生耐药细菌，使人体内产生抗生素抗性基因，致使抗生素对一些炎症及细菌性侵入性疾病无抵抗作用[26]。这些污染物首先对宿主造成毒性作用，如图2-2所示，肠道菌群作为人体的“超级器官”自然也会受污染物的影响而产生变化。重金属暴露会造成肠道菌群的组成、代谢水平及功能发生变化[27]，如砷暴露会使厚壁菌门的丰度显著下降、拟杆菌门的丰富上升，同时胆汁酸代谢和氨基酸代谢被扰乱[28]。抗生素污染会使肠道菌群中的有害菌增多，有益菌如乳酸菌减少，并且使肠道菌群的多样性降低[29]。肠道菌群具有一定的屏障功能，在一定的范围内可对污染物进行抵抗，从而减少进入机体血液及组织器官的污染物，降低污染物对宿主的危害，因此，对肠道菌群与污染物的相互作用值得进一步探究。

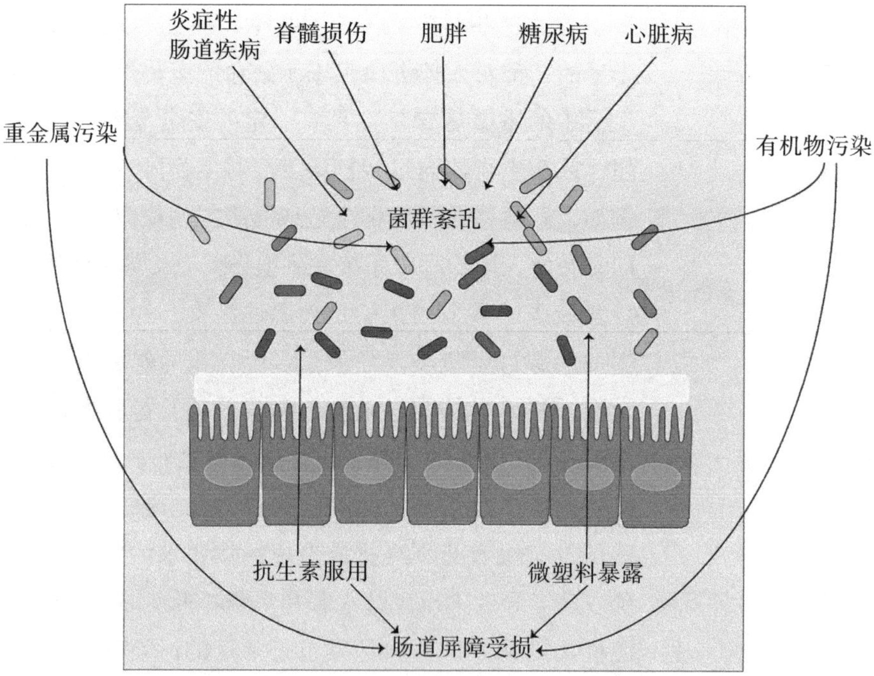

图2-2　疾病、污染物对肠道菌群的影响

2.2　重金属与肠道菌群

常见重金属的质量浓度通常大于5.0 g/cm³，如铬（Cd）、铅（Pb）、镉（Cd）、锌（Zn），汞（Hg）等[30]。近年来随着社会经济的发展，重金属污染在世界各地越来越严重和普遍，中国是全球最大的重金属消耗和产出国之一。农业、工业及矿业生产等诸多方面都有重金属的参与，当然这类污染物对大气层、水体、土壤及人体健康都带来了一定的风险。在对江西省农业区230份土壤表面的重金属进行测定分析后发现，Cu、Zn、Pb、Cr、Ni以及Cd的含量分别是0.48%、0.58%、2.84%、2.41%、0.74%和0.68%，Hg的含量大约为38%[31, 32]。为了测定城市区域土壤重金属含量，来自辽宁鞍山的115份土壤样本被分析，其中污染指数较高的重金属为Cd、Zn、Cu和Pb[31, 33]。水源是人类生活及生产的必需品，重金属对水的污染也是其最常见的污染方式之一，不论是对海水的污染还是对河流的污染，都会造成一定的风险和损害。研究者对长江流域附近的50个不同位置的水体进行了采样，分析了其中8种重金属（As、Cd、Cr、Cu、Ni、Pb、Zn和Hg）的质量浓度，其中Zn的最高（10.42±9.07 μg/L），主要是渔船上的防污涂料造成的[34]。在中国南部排放至珠江口的工业污水每年达到了约两亿吨，其中Cd、Pb、Zn在珠江中的含量达到了惊人的水平，这对生态系统和人类健康带来了巨大的风险[35]。空气中飘浮的灰尘量是评价环境质量的重要指标，灰尘主要来源于工业排放物、燃烧烟尘及土壤扬尘。这些灰尘也是重金属临时的包埋物，从而使重金属在空气中飘浮造成大气层的污染[31]。在对商业街道、学校、景区等地区的灰尘收集分析后，发现Cr、Cu、Zn、Cd和Pb的含量远远高于其在原始土壤中的含量，Ni、Cu、Zn和Pb在景区的含量最高[36]。

人类重金属污染主要有两大来源：一是工业化生产产生，这一类主要集中于从事工业生产的特定人员，由于职业原因长期摄入重金属从而使机体受到严重损害；另一类是日常生活接触产生，如水源污染、土壤污染、食源性污染、大气污染及日常生活用品污染，这类污染物主要通过皮肤、口腔等的微量摄入造成人体的慢性损伤。不论是哪种污染来源最终都会对人体健康造成极大的威胁。As、Cd、Cr、Cu、Ni暴露主要是通过皮肤吸收进入体内，Pb和Zn主要通过消化途径进入体内[37]。Cd暴露会造成肾小管功能障碍，导致肾脏损伤，同时骨骼也会发生病变[38]。三价铬通常毒性较低且是人体内的一种微量元素，其与葡萄糖代谢密切相关，六价铬的毒性是三价铬的100倍，六价铬长期暴露会对呼吸道产生损伤，导致支气管炎及肺功能下降。另外，口腔摄入六价铬会直接造成肝、胃肠道系统和免疫系统的损伤[39]。Pb暴露会直接影响人体神经中枢系统，并造成多器官损害，严重时甚至会造成死亡[40]。以上几种重金属不论是对环境生态系统还是对人体健康都会造成不可逆转的损伤，当然肠道菌群在重金属的暴露中也会受重要的影响，表2-1列举了不同重金属暴露对肠道菌群组成的影响。

表2-1 不同重金属作用下的肠道菌群变化

重金属	肠道菌群变化	作用
Cd	*Akkermansia muciniphila*减少 *Acinetobacter*增多 *Aeromonas hydrophila*减少	结肠炎 内毒素血症 肠道屏障受损
Pb	Bacteroides和Firmicutes下降 Bacteroides增多 *Bacillus desalinosa*减少	炎症反应 肠道屏障受损 耐药菌产生
Cr	*Lactobacillus*增多 Bacteroidetes下降 *Bacillus*增多	具有六价铬的抗性还原为三价铬 — 结合六价铬排出体外
Cu	*Actinomycetes*增多 *Lactobacillus*下降 *Enterobacter*增多 *Coriobacteriaceae*_UCG-002增多	肠道屏障受损,紧密连接蛋白减少 糖代谢受影响 脂代谢受影响 降低抗氧化酶活性
Ni	*Ruminococcus*增多 *Lactobacillus*和*Bifidobacterium*减少	肠道屏障功能受损 肠道菌群多样性减少,功能代谢改变
Hg	*Treponema*增多 *Dehalobacteri*和*Coprococcus*增多	肠道吸收损伤 细菌感染

（1）镉

重金属镉在工业生产中被广泛用于合金制造、电镀等相关领域，镉污染会对人体造成一定的健康威胁。在欧洲，成年人平均每周摄入镉的质量分数大约为2.5 μg/kg[41]。包括镉在内的重金属暴露会发生氧化还原反应，氧化和抗氧化的不平衡会导致机体产生自由基的积累，最终造成氧化应激效应。镉造成氧化应激效应后，会进入细胞造成DNA损伤，从而使细胞死亡。镉是一种危害极大的致癌物，可导致前列腺癌、肺癌、胃肠道癌[42]。肠道菌群作为重要的屏障保护因子，在污染物包括重金属的暴露时具有一定的调节作用，对于肠道菌群缓解镉暴露下的毒性机制尚不清楚。利用无菌小鼠已证实了肠道菌群的存在可以减少机体在镉暴露条件下的负担，与传统的无特异性病原体的小鼠相比，无菌小鼠经镉处理后其血液、器官中镉的含量明显增多，另外粪便中镉的含量是正常小鼠粪便中镉含量的5～30倍，这表明镉要在肠道中持续留存，且这一过程需要复杂的微生物群落参与[43]。银鲫鱼在水中镉暴露条件下，其肠道菌群在属水平中的拟杆菌、阿克曼氏菌、不动杆菌（*Acinetobacter*）以及嗜水气单胞菌（*Aeromonas hydrophila*）均发生了不同水平的变化。其中，阿克曼氏菌是一种肠道益生菌，黏附于黏液层生长，具有保护肠道屏障功能的作用，在镉暴露条件下，阿克曼氏菌的丰度显著下降[44]；不动杆菌的相对丰度却显著上升，研究表明不动杆菌会影响鱼的健康，其相对丰富的增加表明鱼的肠道功能受到损害；嗜水气单胞菌具有黏附能力，可在肠黏膜表面定植，参与宿主营养供应、黏膜防御和免疫，且在健康鱼体内含量较高，可参与消化并与肠炎呈负相关[45，46]。但随着镉暴露的时间及其浓度的增加，嗜水气单胞菌相对丰富显著下降。而在小鼠中镉暴露会明显地降低肠道菌群的丰度及多样性，同时在科水平的毛螺菌科（Lachnospiraceae）的丰度降低，这些都与肠道炎症相关并被认为是结肠炎的主要因素，另外乳酸菌的丰度显著上升，这表明其对镉有一定的耐受性，可减少镉离子在体内的积累[47]。如图2-3所示，镉暴露除了对肠道菌群的组成有影响之外，也会造成肠道微生物组代谢功能的改变，镉暴露使得肠壁受损，从而影响肠道的渗透作用，使得大分子物质进入肠道与肠道菌群相互作用，引起内毒素血症和炎症反应，部分有益菌的下调会减少短链脂肪酸的产生，同时镉诱导的肠系膜损伤本身是微生物受损的一个重要因素[48, 49]。镉诱导的微生物代谢活性受损以及菌群失调会造成其他靶器官的受损。

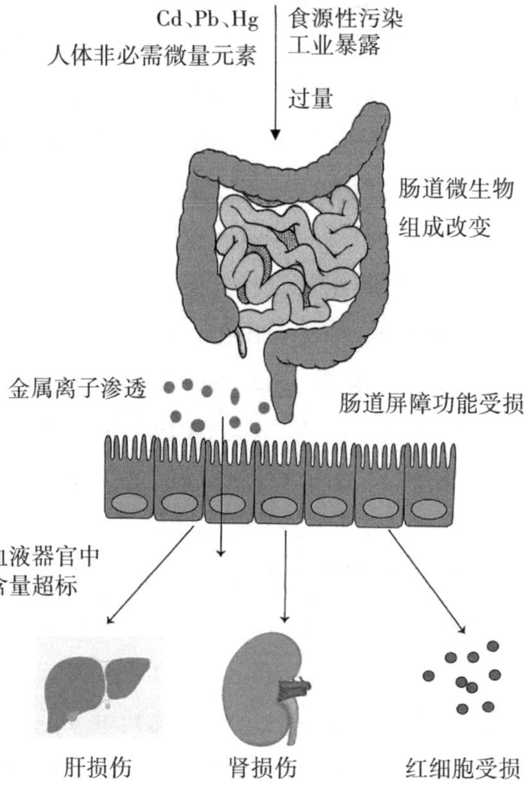

图2-3　镉、铅、汞含量超标对肠道菌群及血液器官的影响

（2）铅

铅是一种对人体健康有严重危害作用的重金属，理想状态下人体铅含量为零。在工业生产中，铅被用于制造蓄电池、电缆，铅也是汽油添加剂，另外铅化合物也被用作塑料和颜料及橡胶的原料[50]。人类接触到的铅污染主要是食源性污染，在日常生活中用的陶瓷碗、茶具以

及方便塑料袋均含有铅。对于人体来说，铅污染是一个慢性中毒的过程，会危及神经系统、消化系统及肝、肾。儿童由于神经系统处于迅速发育时期，更易受铅毒性作用的危害，铅在儿童体内毒性发挥更早且危害更大[51]。肠道菌群在铅暴露的条件下会出现组成和相对丰度的变化。低浓度铅处理小鼠后，发现小鼠肠道菌群的多样性及丰度均发生了变化，拟杆菌门和厚壁菌门的丰度降低，同时在属水平拟杆菌属相对丰度增加，脱盐杆菌属相对丰度降低。拟杆菌属是机会致病菌，此类菌经常参与感染性疾病并且会对抗菌药物产生耐药性。低浓度铅暴露除了会对肠道菌群组成造成影响之外，也会对肠道菌群的代谢带来改变，短链脂肪酸是肠道菌群重要的代谢产物之一，异丁酸盐在铅暴露后浓度明显升高[52]。同样，用铅处理后的鲤鱼，其肠道的完整性受到损伤，对其肠道菌群组成分析发现拟杆菌门增多，同时发现鲤鱼血液中的脂多糖含量增多。产生脂多糖的菌株大部分是拟杆菌，拟杆菌增加机体的炎症反应的同时可导致肠道屏障受损，而在属水平拟杆菌属和毗邻单胞菌属增多，这两种菌都会造成鱼腹泻和肠道炎症，而阿克曼氏菌的显著下降会造成肠道屏障功能损伤，因为阿克曼氏菌具有治疗诸多疾病的能力，如癌症、肥胖、结肠炎等[53，54]。铅的暴露会造成肠道菌群代谢的紊乱，进而影响肠道的屏障功能，同时造成肝和肾损伤，因此铅污染与肠道菌群相互作用的同时会导致组织和器官受威胁，从而使机体中毒。对于铅是如何引起肠道菌群紊乱的，其机制尚不明确。

（3）铬

铬常被用于制作不锈钢、汽车零件、磁带等，是现代科技中重要的金属。自然界中铬的存在主要以六价铬和三价铬为主，且两者可相互转换。六价铬的溶解度高于三价铬，其毒性作用是三价铬的100倍，且易被人体吸收。而三价铬是人体内一种必要的微量金属元素。研究表明，三价铬化合物可维持正常的糖耐量，协助胰岛素在体内维持正常的糖代谢和脂代谢[55]。但六价铬的污染会对人体健康造成损害，如长期摄入六价铬会致癌，同时长期铬的暴露会产生氧化应激效应。六价铬会使体内过氧化氢酶的含量减少，导致过氧化氢含量上升，而过氧化氢会与六价铬结合生成羟自由基[56]。羟自由基和过氧化氢是主要的活性氧（reactive oxygen species，ROS）自由基产物，其可进入细胞内部与蛋白质和DNA结合造成DNA双链断裂及相关酶类失活，从而使细胞发生癌变[57]。肠道菌群是降低六价铬毒性的第一道防线，相关研究证实肠道微生物对六价铬具有一定的耐受力，正常人体摄入1～10 ppm的六价铬都是安全的，不会造成健康威胁，这是由于在肠道菌群的作用下摄入体内的铬被还原成了三价铬[58]。对于较高浓度的铬暴露，肠道菌群可对铬隔绝，以减少六价铬离子进入人体细胞后对蛋白质甚至DNA造成损伤，菌群与铬的相互作用机制在不同的生物样本中并不相同，但是抗性作用和还原性是最普遍最常见的。在肠道菌群与六价铬的相互作用中，将六价铬还原是最有效的缓解毒性的方式，相关研究证实了在肠道菌群中啮齿粪杆菌（*Faecalibaculum rodentium*）编码相关基因产生铬还原酶FcrR，可将六价铬还原为三价铬[59]。当然，部分肠道菌株会对六价铬有一定的抗性，人体粪便中分离获得的韦腊链霉菌LD22（*Streptomyces werraensis* LD22）菌株对$K_2Cr_2O_4$、$NiCl_2$等重金属盐的毒性具有耐受力，乳酸杆菌在六价铬的长时间暴露条件下会产生抗性菌株，该抗性菌株可以与人体免疫系统中的部分细胞一起将铬转换为毒性较低的形式[60，61]。同时部分肠道菌株可与铬离子结合排出体外，革兰氏阳性菌中以芽孢杆菌为主，其细胞壁含有大量肽聚糖和磷酸盐可对铬离子进行吸附，促进铬从细胞内部排出[61]。高浓度或长期六价铬暴露时会对具有抗性和还原能力的菌株产生毒性作用，如图2-4所示，肠道菌群的组成也受到影响，肠道菌群的屏障功能受损进而使机体受到损伤。六价铬的长期暴露会降低鸡肠道菌群的多样性，使拟杆菌门含量下降而厚壁菌门含量上升，同时以拟杆菌属为首的30多种菌的相对丰度下降。肠道菌群的改变不仅会影响肠道自身的功能，还会使整个胃肠道系统发生紊乱，致使影响其他器官组织[62]。

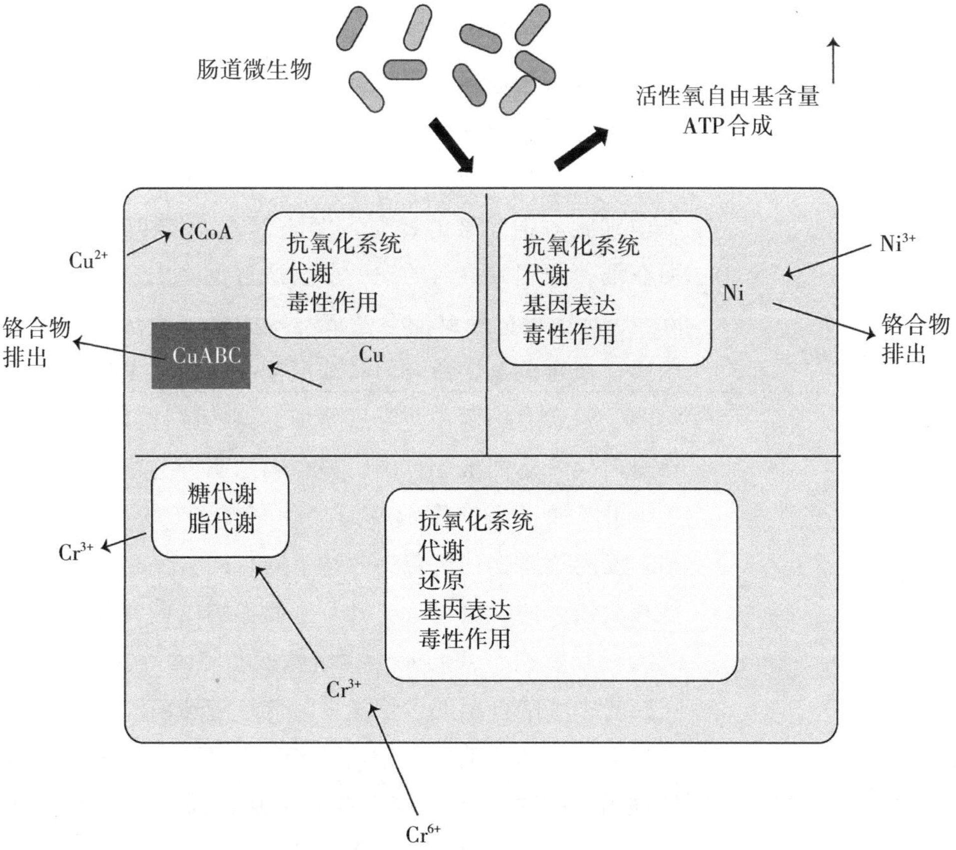

图2-4 铬、铜、镍等人体必需元素在肠道菌群作用下的作用总机制

（4）铜

铜及其合金具有上千年的历史，早在夏商周时青铜器的发展已达到鼎盛时期。铜与人类活动关系密切，铜被广泛应用于机械制造、轻工、国防等领域。当然，铜也是人体内的一种必需微量元素，是许多酶和蛋白的主要组成成分。铜主要起催化作用，细胞色素氧化酶和超氧化物歧化酶（glutathione peroxidase，GPX）均为含铜金属酶。膳食中摄入的铜大部分会被转运至肝参与金属硫蛋白、血浆铜蓝蛋白等的合成[63]。人体日常饮食中摄入的铜大部分会被肠道吸收而后通过特定的转运蛋白运送到肝及其他组织器官。肠道菌群作为重要的免疫保护屏障参与铜在体内的吸收转运。肠道微生物通过跨膜机制对铜进行运输，金属硫蛋白可对铜进行隔绝并且通过氧化降低铜离子的毒性以维持铜的内稳态。铜在进入细菌细胞后，可与金属硫蛋白或者铜结合蛋白结合来维持细菌细胞对铜元素的补充，保证相关代谢和酶的活性，而在细菌细胞内摄入的铜一般会通过铜输出系统进行移除[64]。高浓度的铬暴露会造成一定的机体损伤、胃肠道刺激，甚至危及生命。而高浓度铜会使肠道菌群中的致病菌及机会致病菌相对丰度增加，这些菌株会随着铜的暴露产生一定的抗性，同时整个肠道菌群的组成也会发生变化[65，66]。高浓度铜暴露会使猪肠道的屏障功能受损，使肠道紧密连接蛋白表达下降，同时使小肠中支原体和放线菌的含量显著上升，进而影响戊糖和葡萄糖苷酸的相互转换，并同时干扰组氨酸和嘌呤等相关代谢，在结肠中乳酸杆菌的含量显著降低，链球菌和肠杆菌的含量增多，这影响了甘油磷酸酯的代谢[67]。口腔慢性铜处理后小鼠的肠道菌群多样性下降，同时会影响拟杆菌、疣微菌UCG-014等相关菌属的组成[68]。

（5）镍

镍是良好的合金配方，主要用于工业领域，也用于催化剂制备和货币制造，可电镀在其他金属表面用来防腐。镍也是人体中的一种必需微量元素，主要分布于人体的肺、肝、肾等器官，镍参与人体内某些酶的代谢，相关研究表明镍可激活胰岛素分子降低血糖[69]。镍在细胞中与部分酶活性相关，镍可催化脲酶等金属活性部位的生化功能，多数微生物含有Ni^{2+}结合的活性位点，如红螺菌（*Rhodospirillum*）、金黄色葡萄球菌（*Staphylococcus aureus*）、克雷伯氏菌（*Klebsiella*）、假单胞菌（*Pseudomonas*）、普雷沃氏菌（*Prevotella*）等[70]。

镍在原核和真核细胞中的重要性依赖于Ni^{2+}的酶在肠道菌群中的作用，镍参与宿主代谢以及对相关疾病的响应，因此镍在肠道菌群中既可以激活酶参与代谢，同时也与致病菌的产生有关，如乳酸杆菌可以对金属离子进行结合，保护肠道不受刺激和损伤，Ni^{2+}也可通过这种方式进入机体组织器官，另外双歧杆菌具有脲酶的活性，可与镍进行有效的结合后激活酶活性[71]。当然高浓度的镍暴露也会产生毒性作用，会引起皮肤炎症，冶炼工人长时间接触镍易患鼻癌和肺癌。当然镍的毒性和其形态有关，一般的镍金属盐不具有毒性，但是羰基镍和硫化镍或氯化镍等的毒性较大，人摄入后中枢神经系统等会受损，肾等器官会出现水肿，同时会产生细胞癌变，研究发现白血病患者血液中镍的含量是正常人的2～5倍[72]。这些有毒性的镍盐同样会造成肠道氧化应激效应，并改变肠道菌群的组成。小鼠用1.6 mg/mL的镍暴露后，其肠道中的丙二醛（malondialdehyde，MDA）含量上升，谷胱甘肽过氧化物酶、过氧化氢酶（catalase，CAT）和超氧化物歧化酶（superoxide dismutase，SOD）等抗氧化酶的含量下降，同时在肠道菌群组成上放线菌门含量下降，厚壁菌门含量上升，另外镍处理后在属水平出现红螺菌UCG-002，其可降低抗氧化酶的活性，另外在镍暴露组中瘤胃球菌（*Ruminococcus*）的相对丰度增加，瘤胃球菌增多会降低肠道屏障功能的完整性[73]。利用不同浓度的氯化镍处理肉鸡后发现鸡肠道菌群多样性发生了变化，其中乳酸菌和双歧杆菌的含量显著下降，大肠杆菌和肠球菌的含量增加，并且整个微生物组成的多样性显著下降，这表明氯化镍对肠道菌群及肉鸡具有一定的毒性作用[74]。小鼠口服低剂量硫酸镍后，厚壁菌门和拟杆菌门的比例下降，在属水平拟杆菌和产丁酸肠杆菌的相对丰度上升，厚壁菌门和拟杆菌门的比例下降与免疫反应有关。相关研究表明在过敏反应中拟杆菌含量降低，产丁酸肠杆菌具有抗炎效应，因此在低剂量的镍暴露中其具有一定的耐受性可抵抗镍暴露的毒性作用，而毛螺菌科NK4A136和毛螺菌科UCG-001相对丰度均下降，这两种菌是肠道中的有益菌，可产生大量的丁酸盐和丙酸盐并调控肠道的抗炎效应[75]。

（6）汞

汞是常温状态下唯一以液态呈现的重金属，常见的汞制品为温度计。气态汞可用于制作汞蒸气灯，也可将金从矿中分离。汞是人体的非必需元素，是剧毒性重金属，广泛存在于环境中，遍布全球各个角落。汞在常温下即可蒸发，并以蒸气状态弥散在空气中，再通过人体的呼吸道被吸收经血液循环扩散至全身，可造成中枢神经系统受损，另外无机汞和有机汞化合物均具有不同程度的毒性作用，无机汞通常通过消化道和呼吸道进入人体，可对人体肾、肝等器官造成损伤，而有机汞更易被人体吸收，其中甲基汞的毒性作用最强，可对神经及肾等组织器官造成损伤，其中神经系统受损最为严重[76]。汞易被消化吸收，且元素汞和无机汞在微生物的作用下可转换成有机汞，因此在汞的作用下会对肠道菌群造成影响，如图2-5所示。用80 mg/L的$HgCl_2$对小鼠连续处理90天后发现小鼠的肠道屏障功能受损，并且促进细胞凋亡的基因表达上调，而抑制凋亡的基因表达下调，另外经高通量测序发现，汞暴露会使密螺旋体属（*Treponema*）、脱卤素杆菌属（*Dehalobacterium*）、粪球菌属（*Coprococcus*）的相关丰度上升，其中密螺旋体属会造成肠道的吸收损伤，脱卤素杆菌属和粪球菌属会造成细菌感染，螺杆菌属（*Helicobacterium*）的增加会提高患癌的风险并且链球菌属（*Streptococcus*）的相对丰度上升会产生呼吸系统受损及皮肤疾病[77]。无机汞和有机汞的主要吸收部位均为肠道，无机汞和有机汞都会使肠道微生物紊乱，肠道代谢异常，紧密连接受损并在肠道内发生免疫反应。除了肠道系统发生变化外，被吸收的汞离子会导致肝脏和神经损伤[78]。口服 $HgCl_2$ 会使小鼠肠道菌群厚壁菌门与拟杆菌门的比例增加，此比例发生变化表示肠道菌群处于不健康的状态，通常在疾病及肥胖人群中出现[79]。肠道菌群对甲基汞的一种调节方式是微生物对甲基汞进行去甲基化，使其成为无机汞从而减少对其的吸收，降低其毒性作用[79]。在甲基汞暴露的孕妇粪便中发现阿克曼氏菌相对丰度增加，阿克曼氏菌是一种嗜黏蛋白菌，利用黏液层蛋白为肠道其他微生物提供能量，丰富的阿克曼氏菌会形成强大的屏障，保护肠道及内部组织细胞。因此甲基汞的暴露会使肠道黏液层发挥屏障功能从而减少甲基汞的吸收[80]。

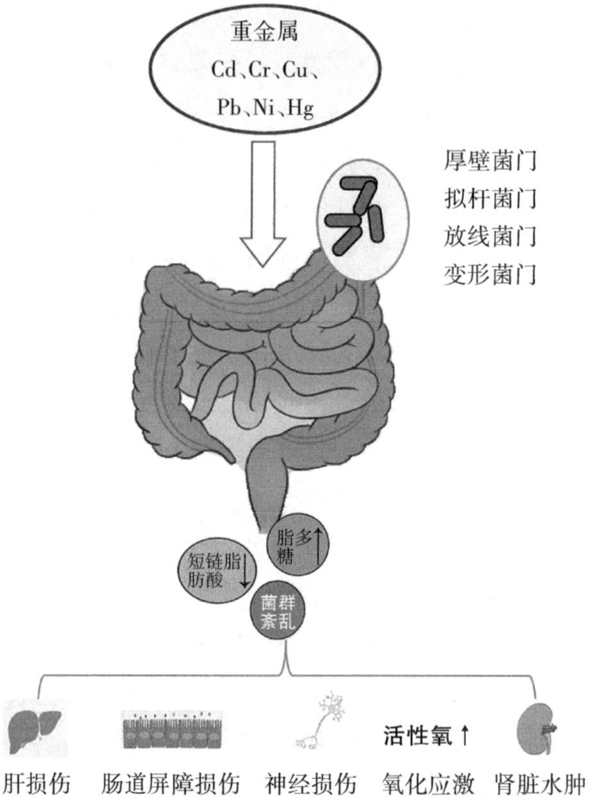

图2-5 重金属暴露对肠道菌群的影响及组织器官的损伤

2.3 持久性有机物与肠道菌群

2.3.1 持久性有机污染物对宿主的危害

持久性有机污染物（persistent organic pollutants，POPs）通常是指化学性质较为稳定，难以降解转换且对人体健康构成严重威胁的有机化学污染物，如多氯联苯、多环芳烃（polycyclic aromatic hydrocarbons，PAHs）等物质。当前环境中持久性有机污染物的暴露途径主要有两种：一种是从事相关生产的职业人员由于长期接触造成持久性有机污染物在体内的慢性积累，另一种是由于环境中的水源突然或长期受持久性有机污染物暴露造成农作物、鱼类虾类以及食用水受持久性有机污染物污染，从而使人体遭受食源性中毒。但是此类污染物易形成气溶胶，大约一周就可以混入大气层中，除了食源性污染及职业暴露外，大气中的持久性有机污染物可通过呼吸道进入人体[81]。持久性有机污染物的毒性作用极大，如图2-6所示，其慢性毒性作用主要是由于有机污染物与脂溶性物质有亲和力，从而使其在脂肪中积累，造成慢性损伤，对肝、肾损伤最大，另外也会对中枢神经系统造成损伤[82]。而急性持久性有机污染物会通过皮肤、消化道、呼吸道进入人体，短期内在人体脂肪组织、肝及肾中蓄积，从而导致机体急性中毒甚至死亡。持久性有机污染物对内分泌系统也会造成不同程度的影响，被认证为潜在的内分泌干扰物质，与雌激素受体有较强的结合能力[83]。装修材料中的低氯多氯联苯可被蒸发出来，教室、公寓等场所中低氯多氯联苯的浓度比户外高出几个数量级。在对低氯多氯联苯展开研究时发现其具有致癌和神经毒性作用，并可能增加人们患糖尿病及心脏病的风险，另外母亲产前接触低氯多氯联苯会间接性地对男性的生殖系统造成损伤，导致男性精子总数下降并增加了患小睾症的风险[84]。多项研究证实乳腺癌的发生风险与多氯联苯、多氟联苯以及类似二噁英的污染物呈正相关[85]。长期暴露接触持久性有机污染物将会增加2型糖尿病的患病风险，进而改变糖代谢并且出现胰岛素抗性，对血液中持久性有机污染物，尤其是有机氯杀虫剂及多氯联苯和胰岛素抗性之间的关系进行了分析，发现它们之间具有一定的关联[86]。大部分持久性有机污染物为脂溶性化合物且具有半挥发性，在宿主中代谢缓慢且能长期积累，另外氯代多氯联苯难以降解，会持久性地对机体造成损伤。肠道上皮组织可吸收持久性有机污染物，这将会造成整个肠道菌群组成的改变，肠道菌群作为有效的保护屏障，具有抵抗外源病原微生物和毒性物质的功能，可降低外源病原微生物和毒性物质对其他组织器官的损伤。各种各样的环境污染物都会对肠道菌群的组成造成改变，并影响整个肠道菌群的内稳态。对于持久性有机污染物的暴露，肠道菌群在一定条件下会与其发生相互作用，进而产生宿主的应激反应。

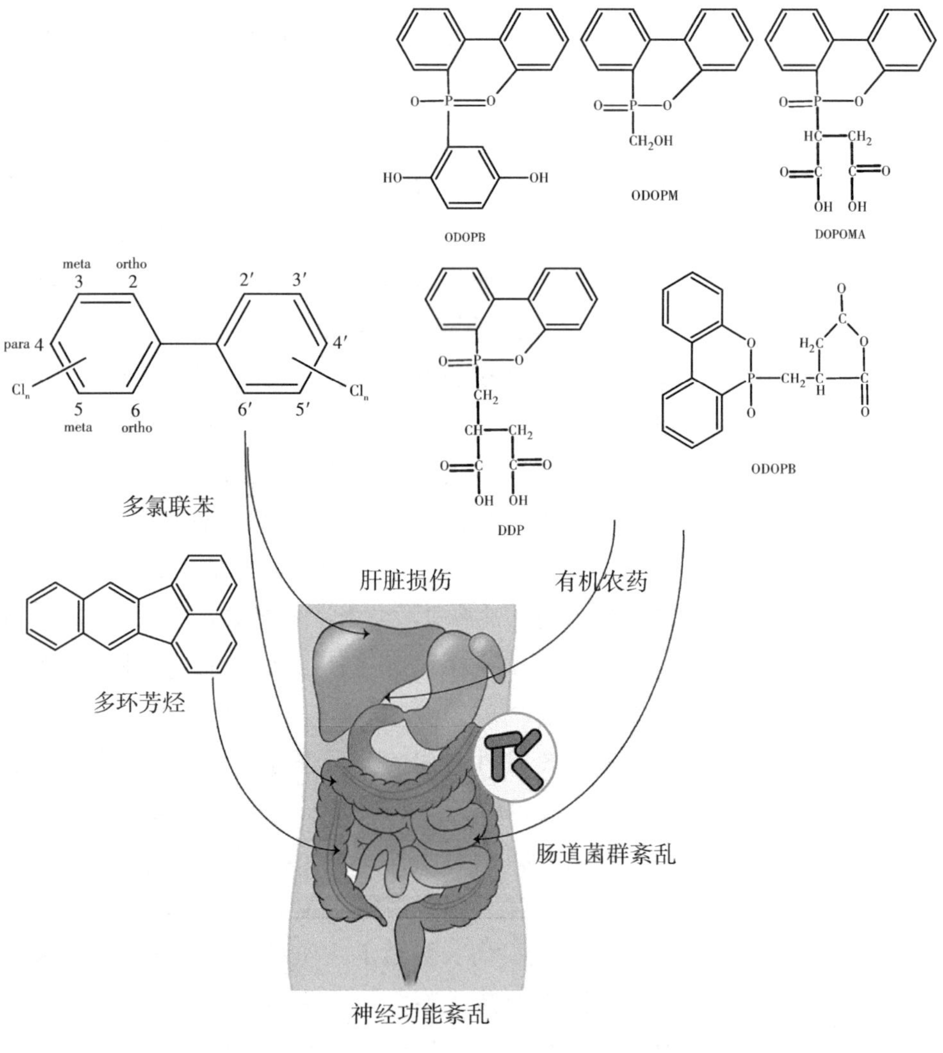

图2-6　持久性有机物暴露与人体健康

2.3.2 不同持久性有机物与肠道菌群的相互作用

（1）多氯联苯

相关研究报道证实持久性有机污染物会造成肠道微生物组成改变进而使肠道菌群的稳态被破坏，同时会影响肠道菌群相关酶活性及代谢，长期的暴露接触必然会使肠道菌群的稳态系统受损[87]。多氯联苯是一种易挥发的有机物，难溶于水但易溶于脂溶性物质，且难以分解，被列为一级致癌物质。在美国普通人群的血清样本中多氯联苯常被检测到，在人类和相关实验动物的研究中发现多氯联苯会造成生殖系统紊乱、发育神经毒性、甲状腺毒性及腹部不适等，对于肠道和多氯联苯作用下的毒性机制尚不清楚，但是在多氯联苯的暴露下肠道菌群的组成发生了改变，其相关的生理功能出现了紊乱。在对小鼠研究中发现，低剂量的多氯联苯会使与初级胆汁酸代谢相关的菌丰度增加，同时肠道菌群的缺乏会促使多氯联苯介导下的与牛磺酸偶联的α和β脂肪酸在肝及大肠内容物中的含量上升，这会使得肠道菌群作用下的由初级胆汁酸合成的内稳态遭受破坏[88]。另外小鼠暴露多氯联苯126会使肠道菌群门和属水平的微生物群向与慢性炎症相关的菌群转移，变形菌门的相对丰度下降，在成年小鼠中口服多氯联苯126会造成肠道菌群多样性下降，同时拟杆菌门与厚壁菌门的比例上升造成肠道内部的炎症反应[89，90]。此外，多氯联苯暴露通过改变乳酸菌、梭状芽孢杆菌、拟杆菌等菌属的胆盐水解酶活性来影响胆汁酸代谢水平[91]。但是，肠道菌群的存在在一定程度上缓解了多氯联苯造成的毒性作用，研究表明多氯联苯会造成小鼠肝脂质积累，并同时出现肝纤维化，小鼠中肠道菌群不受抗生素破坏时可以有效地缓解多氯联苯诱发的肝炎症及纤维化，肠道菌群在此过程中扮演着重要的角色[92]。

（2）多环芳烃

多环芳烃是指含有两个或者两个以上苯环的芳烃，主要通过自然界以及工业生产产生，石油燃烧等都会产生多环芳烃。日常生活用水中已经发现有多环芳烃污染的现象。多环芳烃具有毒性，会造成呼吸系统受损、肝肾损伤，也可引发癌症，被认定为对人体健康有害的污染物之一。人类接触多环芳烃的污染主要源自于空气、食物以及香烟，烟草燃烧产生的多种多环芳烃具有致癌风险。急性多环芳烃中毒通常与其浓度以及和身体接触部位有关，短期多环芳烃暴露会造成哮喘患者肺功能受损并且会让冠心病患者产生血栓。另外，职业人员暴露在高浓度多环芳烃中会出现恶心、腹泻等症状，实验动物长期暴露于低剂量多环芳烃中会出现生殖发育受损[93]。在实验动物中研究发现，多环芳烃会影响肠道微生物的组成，同时儿童遭受多环芳烃暴露会出现神经发育障碍，对38名3岁儿童在遭受多环芳烃暴露后的肠道菌群组成及神经发育进行分析发现，肠道菌群中罗斯菌、瘤胃球菌、布劳特氏菌（*Blautia*）等与丙酸、丁酸产生相关的菌群与多环芳烃暴露呈负相关，肠道菌群的代谢受到影响，另外发现乳酸杆菌与多环芳烃暴露呈正相关，这是由于乳酸杆菌已被研究发现可产生乙酰胆碱，其可以有效地激发乙酰胆碱受体在体内的活性，从而促进人体神经传递[94]。肠道菌群在多环芳烃暴露的条件下具有一定的调节缓冲作用，可减少多环芳烃带来的消极作用，尽力维护肠道系统的内稳态。当然，高剂量的多环芳烃会造成肠道菌群组成改变，让耐受菌增长占据优势，而部分有益菌因毒性作用而死亡，机体进而产生损伤。对暴露在多环芳烃下的大西洋一鳃鱼进行研究发现，鱼的肠道微生物多样性及丰富度下降，同时科水平脱硫杆菌科（Desulfobacteraceae）和属水平螺杆菌属相对丰度显著上升[95]。

（3）有机农药

有机农药是所有有机化合物农药的总称，常见的有有机氯农药、有机磷农药、有机硫农药、酚类化合物等，此类化合物可对农业生产中的害虫、杂草等对农作物有害的生物进行防治。但是残余的有机农药会对土壤、大气、水体产生污染，进而危害人体健康。有机农药急性中毒会使人在短期内出现恶心、晕厥、呼吸困难等症状，往往造成急性中毒症死亡，而在长期慢性中毒的过程中，有机农药会通过呼吸道、消化道、皮肤进入人体，此类有机化合物会与体内的脂肪相溶从而在体内蓄积，抑制胆碱酯酶的活性，造成乙酰胆碱积累从而导致机体的神经功能紊乱[96]。肠道菌群作为第一保护屏障，在有机农药进入体内时会与其接触，参与调节有机农药带来的毒性作用，如图2-7所示。小鼠暴露于杀虫剂草甘膦时其整个肠道菌群的拟杆菌门和厚壁菌门丰度相较于对照组出现差异，其中放线菌门的棒状杆菌，厚壁菌门的乳酸杆菌、拟杆菌相对丰度均出现下降，另外小鼠十二指肠和空肠的绒毛结构发生形态学改变[97]。有爪蟾蜍暴露于顺式联苯菊酯杀虫剂时，发现其会出现肠道菌群组成紊乱，其中拟杆菌丰度显著降低，气单胞菌属（*Aeromonas*）和毗邻单胞菌属等致病菌的含量显著上升，同时伴随着非酒精性脂肪肝的产生，另外在杀虫剂处理组发现肠道内容物中牛磺酸、甘氨胆酸、鹅脱氧胆酰甘氨酸含量显著低于对照组，其代谢水平显著降低[98]。合成除虫菊酯已被证实可抑制革兰氏阴性菌和革兰氏阳性菌的活性，研究发现小鼠暴露于合成除虫菊酯时，其体内的双歧杆菌等有益菌的生长被抑制，此外草甘膦会抑制肠道菌群合成5-烯醇丙酮基3-磷酸盐，同时会影响有益菌的生长，但是沙门氏菌和梭状芽孢杆菌则具有一定的抗性作用[99，100]。

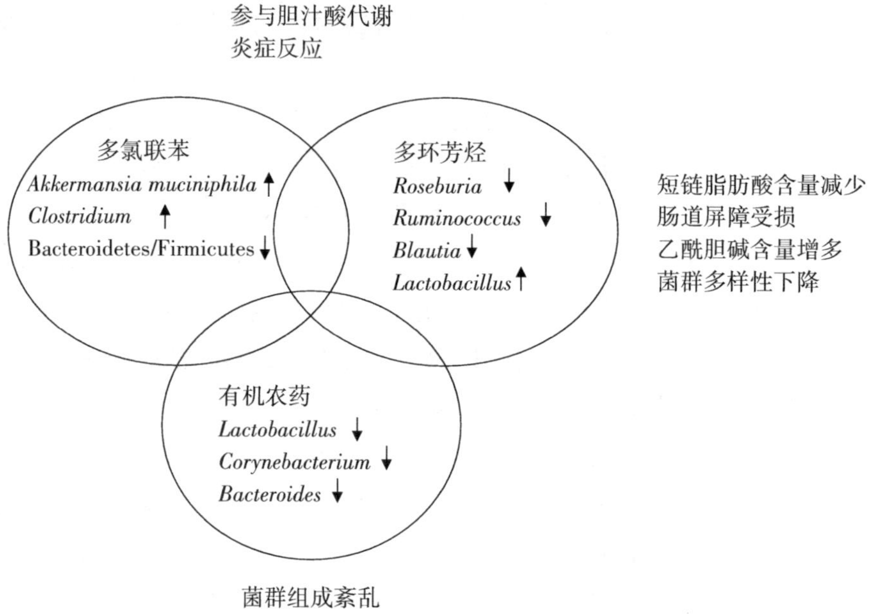

图2-7 持久性有机物暴露引起的肠道菌群变化及机体反应

2.4 抗生素与肠道菌群

2.4.1 抗生素滥用对宿主的危害

使用抗生素是近代科学发展后治疗细菌及病原微生物感染的一种重要医疗手段，20世纪30年代青霉素的成功提取为世界大战的伤员提供了及时的救治机会，青霉素是当时战场上必备的医用物资，挽救了无数人的性命。随着研究的进一步发展，多种抗生素被成功提取，链霉素、四环素、庆大霉素等相继问世，这些抗生素都是从微生物或者高等动植物产生的次级代谢物中提取获得，经研究发现其具有抗病原微生物、抗肿瘤等功效，因而被称为抗生素。但是随着对抗生素的进一步研究发现，抗生素的不合理使用会对人体产生一定的危害。中国是抗生素使用和生产大国，抗生素被广泛地使用在人和动物身上，残留在食物中抗生素会被人体摄入[101]。长期的抗生素摄入会对人体健康造成一定威胁，如造成消化系统功能紊乱、过敏反应，长期低浓度的抗生素暴露具有慢性毒性已被证实[102]。食品中残留的抗生素被长期摄入会致癌、致畸，引发生殖损伤，甚至肥胖[103]。当然抗生素滥用的致命危害是抗生素抗性基因的产生，进而在人体内产生耐药菌。环境中耐药菌的存在会加重各种病原菌的感染，在巴西东南部的工厂和医院中分离获得了两株具有四环素抗性的铜绿假单胞菌——EW32和EW33，这两株耐药菌的存在加快了环境细菌和致病菌之间耐药基因的交换，这给动物和人体感染各种病原菌的治疗增加了困难[104]。

2.4.2 耐药菌的产生

抗生素的大量使用会使病原微生物对抗生素产生一定的耐药性，这类菌通常被称为耐药菌。耐药菌的出现会造成肠道菌群组成紊乱，其会在体内大量繁殖产生致病效应，对人体健康构成致命威胁。抗生素的耐药性可分为四种不同的类型：自然耐药性、获得性抗性、抗性转移和多重耐药性。其中，自然耐药性主要是通过细菌自身的结构介导，使得抗生素无法通过特定的结构对微生物造成损伤；获得性抗性是通过染色体或者额外的染色体结构突变产生，可通过物理化学方式发生；抗性转移是指微生物对特定的抗生素具有抗性的同时对与其具有相同结构的抗生素也具有抗性；多重耐药性是指对抗生素产生耐药性的病原体不被单一的抗生素治疗，因抗生素的多重使用从而产生多种耐药性菌[105, 106]。在对九龙河进行抗生素污染检测时，分析了常见的22种抗生素浓度，其中磺胺类、喹诺酮类和氯霉素类药物被频繁检测到，同时分离出35株菌并分析其耐药性时发现97.1%的细菌具有耐药性，且70%的细菌为多重耐药菌[107]。

2.4.3 抗生素与肠道菌群的相互作用

肠道菌群对于机体生长发育至关重要，在肠道内部的微生物群形成了一个特殊的生态环境维持正常的生理活动。如图2-8所示，抗生素的摄入会对肠道菌群的组成造成影响，相关研究已证实抗生素会扰乱肠道菌群的组成并同时干扰菌群间的相互作用。在小鼠的研究中发现抗生素的摄入会在肠道中增加游离的唾液酸，可以被沙门氏菌和艰难梭状芽孢杆菌等致病菌利用并促进其生长[108]。克林霉素会对肠道菌群的组成产生持久性的影响，在对连续7天服用克林霉素和甲硝唑患者的粪便微生物组成进行分析时，发现克林霉素在暴露后只有部分菌群可快速恢复，且其影响可长达四年之久[109]。对环丙沙星的研究证实其可对肠道菌群的组成产生影响，环丙沙星摄入5天后可发现肠道中三分之一的菌群受影响并且整个肠道菌群的相对丰度显著降低[110]。在用环丙沙星对小鼠处理后，对其粪便菌群培养发现肠道菌群在体外培养时易受沙门氏菌的侵染，整个肠道菌群的组成发生了变化，部分菌的最小抑菌浓度（minimal inhibi concentration，MIC）增大，其中毛螺菌科的MIC最高，说明部分肠道菌群对抗生素出现了耐药性[111]。抗生素处理对肠道菌群影

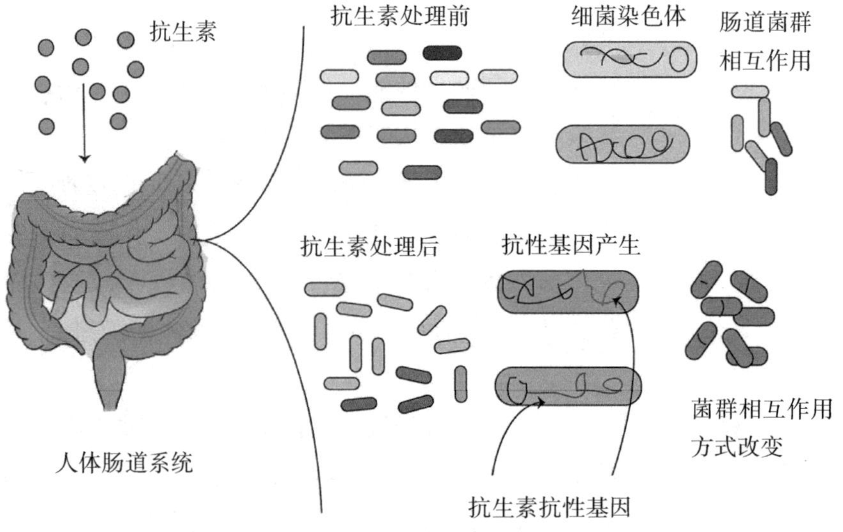

图2-8　抗生素处理前后肠道菌群多样性、染色体结构及菌群相互作用方式的变化

响的关键是致使肠道菌群多样性下降，在成人中使用时，发现万古霉素、庆大霉素会增加大肠杆菌等致病菌的丰度，降低双歧杆菌等有益菌的丰度，整个肠道菌群的组成会在1.5个月内恢复，但是部分菌仍会出现检测不到的现象[112, 113]。当然抗生素处理会改变肠道菌群的代谢，减少毛螺菌科和瘤胃菌属，此类菌属可有效地将阿拉伯糖醇转换为戊糖从而被代谢利用[112]。

2.5 微塑料与肠道菌群

2.5.1 微塑料污染对宿主的危害

微塑料通常是指直径小于5 mm的塑料颗粒，此类污染物因不能溶于水可长期在环境中存在，是国际上最新关注的四大新型污染物之一。目前主要的微塑料污染物有生活垃圾污染物，如废弃的口罩、食用餐盒、吸管、儿童玩具等。此类污染物体积小，比表面积大，在环境中可吸附承载更多的污染物，且不易被降解。有关微塑料对人体健康危害的探究，如图2-9所示，表明微塑料可通过呼吸系统或者口腔进入人体，无论哪种方式摄入的微塑料均会对人体健康造成影响。空气中飘浮的小颗粒微塑料，如直径小于2.5 mm的柴油尾气颗粒，可穿越细胞膜诱发氧化应激和炎症反应，并会诱发呼吸系统疾病、心脑血管疾病及肺癌的风险[114]。对人体细胞、水生动物和小鼠的研究表明，直径小于10 μm的微塑料颗粒可进入肠到淋巴和循环系统，并在全身和各个器官积累，包括肝肾及大脑。在人体细胞体外培养及啮齿动物微塑料暴露研究中发现摄入的微塑料会造成细胞内部氧化应激效应，产生细胞因子，造成细胞损伤进而引发炎症反应和免疫反应，使细胞内部DNA损伤并产生神经毒性和代谢效应[115]。在对8位身体健康的33～65岁的志愿者粪便中微塑料含量检测时发现8份粪便样本均为阳性，其中每10 g样本中含有20个微塑料，50～500 μm不等，此类微塑料均为日常生活中人体无意摄入的[116]。人体中的微塑料主要通过消化、呼吸、皮肤接触等途径摄入，其中日常食物的暴露摄入是常见并且不可避免的一种摄入方式。微塑料进入胃肠道系统可通过食物包裹或者在吸入后可通过呼吸道黏液和纤毛的清除进入肠道，这会导致炎症反应并且增加肠道系统的渗透作用，进而改变肠道微生物的代谢和组成。对人体每天从空气中摄入的微塑料研究发现，一个正常的男性每天通过呼吸摄入微塑料高达272个，并且密度较小且颗粒更小的微塑料进入人体肺部的更深处，巨噬细胞对微塑料进行清除或者将其迁移至淋巴循环的过程中更易异位，进而会引发呼吸系统的慢性炎症[117]。对炎症性肠道疾病患者粪便和健康人粪便进行微塑料含量测定发现，炎症性肠道疾病患者粪便中微塑料浓度远远高于健康人粪便中微塑料的浓度，并且在

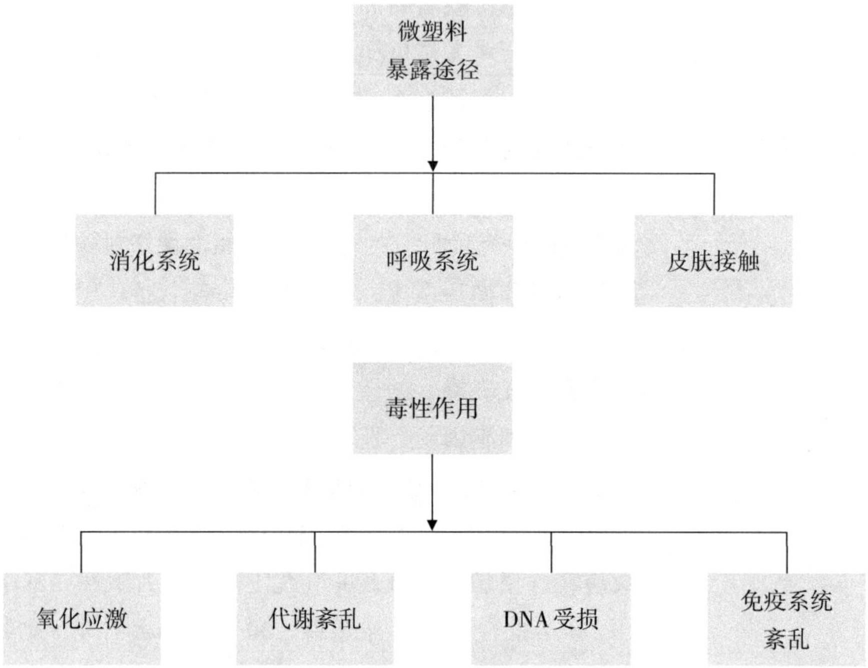

图2-9　微塑料暴露途径及其对宿主的毒性危害

粪便中发现了15种不同种类的微塑料，其中聚酯纤维和聚酰胺的含量最高，微塑料浓度与炎症性肠道疾病呈正相关[118]。同样，摄入的微塑料也不可避免地在人体肺部积累，尸检人体肺部发现29个组织样本中有13个样本含有聚合物和纤维塑料颗粒，并且所有聚合颗粒的直径均小于5.5 μm，纤维颗粒的直径为8.2～16.8 μm[119]。

2.5.2 微塑料与肠道菌群的相互作用

微塑料可通过胃肠道进入人体，这会使其与肠道微生物发生相互作用。肠道菌群在污染物的毒性作用中扮演着重要的角色，是毒理学评价的重要指标，微塑料可进入肠道细胞并对肠道细胞造成损伤，如图2-10所示。此外相关研究报道土壤中微塑料暴露会减少土壤生物肠道菌群的多样性，在跳虫肠道菌群中微塑料暴露会减少拟杆菌门的丰度并且增加厚壁菌门的丰度。大小不同的微塑料会使小鼠肠道菌群失调，降低肠道菌群的多样性并且会影响宿主代谢并造成生理指标异常[120]。微塑料影响肠道菌群的直接原因是微塑料可进入肠道系统并被吸收，进而产生一系列的应激反应。微塑料可通过改变肠道菌群组成来影响宿主的健康，这是未来对毒理学评价的一个新的目标。将雄性小鼠暴露在两种大小不一的微塑料中，5周后发现两种微塑料均可降低小鼠体重以及肝和脂肪的重量，并且两种微塑料均可以减少肠道黏液的含量，在门水平两种微塑料处理后会减少变形菌门和厚壁菌门的相对丰度，厚壁菌门主要参与营养物质的吸收转换，并且与肥胖相关。另外，在盲肠内容物中发现颤螺菌属（*Oscillospira*）和厌氧链球菌（*Anaerostipes*）的含量减少，但普雷沃氏菌和瘤胃球菌等的含量增加，其中颤螺菌属与减重呈正相关并且参与次级胆汁酸的产生，但是普雷沃氏菌与类风湿性关节炎的发病机制有关，另外瘤胃球菌与病态的肥胖有关[121]。在C57BL/6小鼠模型中高浓度的微塑料暴露后小鼠肠道菌群丰度和多样性增加，其中葡萄球菌的相对丰度增加，而副拟杆菌属（*Parabacteroides*）的相对丰度相较于对照组显著降低，在摄入微塑料后肠道菌群的内稳态体系被打破，葡萄球菌的增多会造成肠道内部炎症的发生，在溃疡性结肠炎和肠易激综合征（irritable bowel syndrome，IBS）患者中发现副拟杆菌的含量显著低于正常人，另外微塑料处理后会降低免疫细胞的含量[122]。利用体外胃肠道模型分析聚对苯二甲酸乙二醇酯（polyethylene terephthalate，PET）对人体肠道菌群组成的影响，发现结肠微生物群落组成发生了改变，其中一些微生物群落可以黏附在PET塑料表面，形成生物膜，并且双歧杆菌、梭状芽孢杆菌、总厌氧菌和总好氧菌含量均出现了下降的趋势[123]。在用直径为5 μm的微塑料对小鼠处理90天后，分析发现小鼠

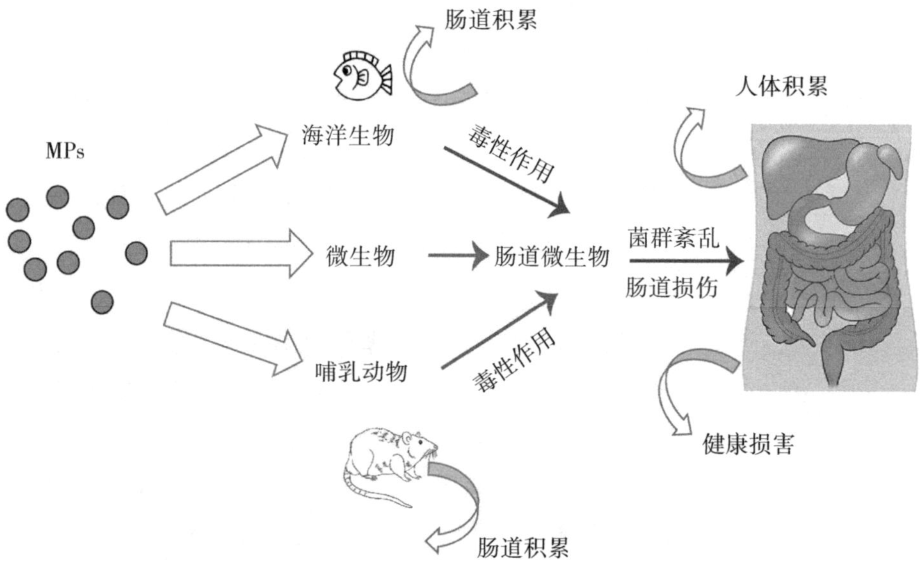

图2-10　微塑料与海洋生物、哺乳动物肠道菌群的相互作用

出现了组织损伤、生殖系统紊乱以及激素合成受抑制，再对肠道菌群分析发现普雷沃氏菌UCG-001与睾丸损伤呈正相关，并且一些促炎细菌的含量会显著上升造成炎症反应[124]。

（王星 令桢民*）

参考文献

[1]RAMAKRISHNA B S. The normal bacterial flora of the human intestine and its regulation[J]. Journal of clinical gastroenterology, 2007, 41(5): 2–6.

[2]JANDHYALA S M, TALUKDAR R, SUBRAMANYAM C, et al. Role of the normal gut microbiota [J]. World journal of gastroenterology, 2015, 21 (29) : 8787–8803.

[3]MARIAT D, FIRMESSE O, LEVENEZ F, et al. The Firmicutes/Bacteroidetes ratio of the human microbiota changes with age[J]. BMC microbiology, 2009(9): 1–6.

[4]SANTOS-MARCOS J A, PEREZ-JIMENEZ F, CAMARGO A. The role of diet and intestinal microbiota in the development of metabolic syndrome[J]. Journal of nutritional biochemistry, 2019(70): 1–27.

[5]ADAK A, KHAN M R. An insight into gut microbiota and its functionalities [J]. Cellular and molecular life sciences, 2019, 76(3): 473–493.

[6]HILL D A, ARTIS D. Intestinal bacteria and the regulation of immune cell homeostasis[J]. Annual review of immunology, 2010(28): 623–667.

[7]SCHWERBROCK N M J, MAKKINK M K, VAN DER SLUIS M, et al. Interleukin 10-deficient mice exhibit defective colonic Muc2 synthesis before and after induction of colitis by commensal bacteria [J]. Inflammatory bowel diseases, 2004, 10 (6): 811–823.

[8] KELLY D, CONWAY S, AMIONV R. Commensal gut bacteria: mechanisms of immune modulation[J]. Trends in immunology, 2005, 26(6): 326–333.

[9]ROOKS M G, GARRETT W S. Gut microbiota, metabolites and host immunity [J]. Nature reviews immunology, 2016, 16(6): 341–352.

[10]FLINT H J, SCOTT K P, LOUIS P. The role of the gut microbiota in nutrition

and health[J]. Nature reviews gastroenterology & hepatology, 2012, 9(10): 577-589.

[11]ZHAO L Y, ZHANG X, ZUO T. The composition of colonic commensal bacteria according to anatomical localization in colorectal cancer[J]. Engineering, 2017, 3(1): 90-97.

[12]KABAT A M, SRINIVASAN N, MALOY K J. Modulation of immune development and function by intestinal microbiota[J]. Trends in immunology, 2014, 35(11): 507-517.

[13]CARDING S, VERBERB E K, VIPOND D T, et al. Dysbiosis of the gut microbiota in disease[J]. Microbiology ecology health disease, 2015, 26(1): 1-9.

[14] HANSEN J, GULATI A, SARTOR R B. The role of mucosal immunity and host genetics in defining intestinal commensal bacteria[J]. Current opinion in gastroenterology, 2010, 26(6): 564-571.

[15]GUNGOR B, ADIGUZEL E, GURSEL I, et al. Intestinal microbiota in patients with spinal cord injury[J]. PLoS one, 2016, 11 (1): 1-10

[16]KIGERL K A, MOSTACADA K, POPPOVICH P G. Gut microbiota are disease-modifying factors after traumatic spinal cord injury[J]. Neurotherapeutics, 2018, 15(1): 60-67.

[17]LIU R X, HONG J, XU X Q, et al. Gut microbiome and serum metabolome alterations in obesity and after weight-loss intervention[J]. Nature medicine, 2017, 23(7): 859-868.

[18]FAN Y, PEDERSEN O. Gut microbiota in human metabolic health and disease[J]. Nature reviews microbiology, 2021, 19(1): 55-71.

[19]ZHONG H, REN H, LU Y, et al. Distinct gut metagenomics and metaproteomics signatures in prediabetics and treatment-naïve type 2 diabetics[J]. EBioMedicine, 2019(47): 373-383.

[20]KARLSSON F H, TREMAROIL V, NOOKAEW I, et al. Gut metagenome in European women with normal, impaired and diabetic glucose control[J]. Nature, 2013, 498(7452): 99-103.

[21]KZAEMIAN N, MAHMOUDI M, HALPERIN F, et al. Gut microbiota and cardiovascular disease: opportunities and challenges[J]. Microbiome, 2020, 8 (1): 1-17 .

[22]TUMOLO M, ANCONA V, DE PAOLA D, et al. Chromium pollution in Euro-

pean water, sources, health risk, and remediation strategies: an overview[J]. International journal of environmental research and public health, 2020, 17 (15): 1-24.

[23] BERNHOFT R A. Mercury toxicity and treatment: a review of the literature [J]. Journal of environmental and public health, 2012(2012): 1-11.

[24]石意. 急诊早期应用呼吸机抢救急性重症有机磷农药中毒致呼吸衰竭的临床表现[J]. 中国医药指南. 2017, 15(17): 43.

[25]景美琪, 李绰然, 王隆清, 等. 微塑料的毒理学研究进展——微塑料对微生物、藻类、鱼类和哺乳动物类的毒理学效应[J]. 生态毒理学报, 2022(4): 83-98.

[26] TELLO A, AUSTIN B, TELFER T C. Selective pressure of antibiotic pollution on bacteria of importance to public health[J]. Environmental health perspectives, 2012, 120(8): 1100-1106.

[27] GAO B, CHI L, MAHBUB R, et al. Multi-omics reveals that lead exposure disturbs gut microbiome development, key metabolites, and metabolic pathways [J]. Chemical research in toxicology, 2017, 30(4): 996-1005.

[28] LU K, ABO R P, SCHLIEPER K A, et al. Arsenic exposure perturbs the gut microbiome and its metabolic profile in mice: an integrated metagenomics and metabolomics analysis[J]. Environmental health perspectives, 2014, 122(3): 284-291.

[29] MODI S R, COKKINS J J, RELMAN D A. Antibiotics and the gut microbiota [J]. Journal of clinical investigation, 2014, 124(10): 4212-4218.

[30] OVES M, KHAN M S, ZAIDI A, et al. Soil contamination, nutritive value, and human health risk assessment of heavy metals: an overview[M]. Berlin: Springer Vienna, 2012.

[31] FEI J C, MIN X B, WANG Z X, et al. Health and ecological risk assessment of heavy metals pollution in an antimony mining region: a case study from South China [J]. Environmental science and pollution research, 2017, 24(35): 27573-27586.

[32] LIU G N, YU Y J, HOU J. et al. An ecological risk assessment of heavy metal pollution of the agricultural ecosystem near a lead-acid battery factory[J]. Ecological Indicators, 2014(47): 210-218.

[33] REN W, XUE B, GENG Y, et al. Inventorying heavy metal pollution in redeveloped brownfield and its policy contribution: case study from Tiexi District, Shenyang, China[J]. Land use policy, 2014(38): 138-146.

[34]YIN S,FENG C H,LI Y Y,et al. Heavy metal pollution in the surface water of the Yangtze Estuary:a 5-year follow-up study[J]. Chemosphere,2015(138):718-725.

[35]CHEN X Y,GAO H W,YAO X H,et al. Ecosystem health assessment in the pearl river estuary of China by considering ecosystem coordination[J]. PLoS one,2017,8 (7):1-10

[36]WEI X,GAO B,WANG P,et al. Pollution characteristics and health risk assessment of heavy metals in street dusts from different functional areas in Beijing,China [J]. Ecotoxicology and environmental safety,2015(112):186-192.

[37]LI Z Y,MA Z W,VAN DER KUIJP T J,et al. A review of soil heavy metal pollution from mines in China:pollution and health risk assessment[J]. Science of the total environment,2014(468):843-853.

[38] ELINDER C G,JARUP L. Cadmium exposure and health risks:recent findings[J]. Ambio,1996,25(5):370-373.

[39]SAHA R,NANDI R,SAHA B. Sources and toxicity of hexavalent chromium [J]. Journal of coordination chemistry,2011,64(10):1782-1806.

[40]DEBNATH B,SINGH W S,MANNA K. Sources and toxicological effects of lead on human health[J]. Indian journal of medical specialities,2019,10(2):1-6.

[41]NAWROT T S,STAESSEN J A,ROELS H A,et al. Cadmium exposure in the population from health risks to strategies of prevention[J]. Biometals,2010,23(5):769-782.

[42]JOMOVA K,VALKO M. Advances in metal-induced oxidative stress and human disease[J]. Toxicology,2011,283(2/3):65-87.

[43]BRETON J,DANIEL C,DEWULF J,et al. Gut microbiota limits heavy metals burden caused by chronic oral exposure[J]. Toxicology letters,2013,222(2):132-138.

[44] WANG N,GUO Z,ZHANG Y,et al. Effect on intestinal microbiota,bioaccumulation,and oxidative stress of Carassius auratus gibelio under waterborne cadmium exposure[J]. Fish physiology and biochemistry,2020,46(6):2299-2309.

[45]NAVARRETE P,FUENTES P,DE LA FUENTE L,et al. Short-term effects of dietary soybean meal and lactic acid bacteria on the intestinal morphology and micro-

biota of A tlantic salmon (S almo salar)[J]. Aquaculture nutrition,2013,19(5):827-836.

[46]NAMBA A,MANO N,HIROSE H. Phylogenetic analysis of intestinal bacteria and their adhesive capability in relation to the intestinal mucus of carp[J]. Journal of applied microbiology,2007,102(5):1307-1317.

[47]MASSOUD R,ZOGHI A. Potential probiotic strains with heavy metals and mycotoxins bioremoval capacity for application in foodstuffs[J]. Journal of applied microbiolog,2022,133 (3):1288-1307

[48]JOLY CONDETTE C,KHORSI-CAUTE H,MORKIERE P,et al. Increased gut permeability and bacterial translocation after chronic chlorpyrifos exposure in rats [J]. PLoS one,2014,9(7):1-10.

[49]TINKOV A A,GRITSENKO V A,SKALNAYA M G,et al. Gut as a target for cadmium toxicity[J]. Environment pollution,2018(235):429-434.

[50]梁奇峰,李京雄,丘基祥. 环境铅污染与人体健康[J].广东微量元素科学,2003,10(7):57-60.

[51]杨田,王文瑞. 食品中铅污染与人体健康[J]. 世界最新医学信息文摘,2014,14(7):44-46.

[52]XIA J,JIN C,PAN Z,et al. Chronic exposure to low concentrations of lead induces metabolic disorder and dysbiosis of the gut microbiota in mice[J]. Science of the total environment,2018(631):439-448.

[53]ZHANG Y,ZHANF P,SHANG X,et al. Exposure of lead on intestinal structural integrity and the diversity of gut microbiota of common carp[J]. Comparative biochemistry and 1physiology C-toxicology & pharmacology,2021(239):1-7.

[54]YOUMANS B P,AJAMI N J,JIANG Z D,et al. Characterization of the human gut microbiome during travelers' diarrhea[J]. Gut microbes,2015,6(2):110-119.

[55]VINCENT J B. Recent advances in the nutritional biochemistry of trivalent chromium[J]. Proceedings of the nutrition society,2004,63(1):41-47.

[56]AHMAD M K,SYMA S,MAHMOOD R. Cr(Ⅵ)induces lipid peroxidation, protein oxidation and alters the activities of antioxidant enzymes in human erythrocytes [J]. Biological trace element research,2011,144(1):426-435.

[57]BORTHIRY G R,ANYHOLINE W E,KALYANARAMAN B,et al. Reduc-

tion of hexavalent chromium by human cytochrome b5: generation of hydroxyl radical and superoxide[J]. Free radical biology and medicine, 2007, 42(6): 738–755.

[58]SHRIVASTAVA R, UPRETI R K, CHATURVEDI U C. Various cells of the immune system and intestine differ in their capacity to reduce hexavalent chromium[J]. FEMS immunology & medical microbiology, 2003, 38(1): 65–70.

[59]FENG P, YE Z, HAN H, et al. Tibet plateau probiotic mitigates chromate toxicity in mice by alleviating oxidative stress in gut microbiota[J]. Community biology, 2020, 3(1): 1–12.

[60]LATHA S, VINOTHINI G, DHANASEKARAN D. Chromium [Cr(Ⅵ)] biosorption property of the newly isolated actinobacterial probiont *Streptomyces werraensis* LD22[J]. 3 Biotech, 2015, 5(4): 423–432.

[61]YOUNAN S, SAKITA GZ, ALBUQUERQUE T R, et al. Chromium(Ⅵ) bioremediation by probiotics[J]. Journal of the science of food and agriculture, 2016, 96(12): 3977–3982.

[62]LI A, DING J, SHEN T, et al. Environmental hexavalent chromium exposure induces gut microbial dysbiosis in chickens[J]. Ecotoxicology environment safety, 2021(227): 1–11.

[63]付鹏钰，韩涵，叶冰，等. 微量元素铜对人体健康的影响[J]. 河南预防医学杂志，2021，32(12): 888–892.

[64]ANDREI A, ÖZTURK Y, KHALFAOUI-HASSANI B, et al. Cu homeostasis in bacteria: the ins and outs[J]. Membranes, 2020, 10(9): 1–46.

[65]ZHANG R, GU J, WANG X, et al. Response of antibiotic resistance genes abundance by graphene oxide during the anaerobic digestion of swine manure with copper pollution[J]. Science of the total environment, 2019(654): 292–299.

[66]PAJARILLO E A B, LEE E, KANG D K. Trace metals and animal health: interplay of the gut microbiota with iron, manganese, zinc, and copper[J]. Animal nutrition, 2021, 7(3): 750–761.

[67]LIAO J, LI Q, LEI C, et al. Toxic effects of copper on the jejunum and colon of pigs: mechanisms related to gut barrier dysfunction and inflammation influenced by the gut microbiota[J]. Food function, 2021, 12(20): 9642–9657.

[68]ZHAI Q, LI T, YU L, et al. Effects of subchronic oral toxic metal exposure on

the intestinal microbiota of mice[J]. Science bulletin,2017,62(12):831-840.

[69] 吴茂江. 镍与人体健康[J]. 微量元素与健康研究,2014,31(1):74-75.

[70]GIMANBO F,ITALIA S,TEODORO M,et al. Influence of toxic metal exposure on the gut microbiota[J]. World academy of sciences journal. 2021, 18 (11): 5510-5510.

[71]ZAMBELLI B,UVERSKY V N,CIRULI S. Nickel impact on human health: an intrinsic disorder perspective[J]. Biochimica et biophysica acta,2016,1864(12): 1714-1731.

[72]韦友欢,黄秋婵,苏秀芳. 镍对人体健康的危害效应及其机理[J].环境科学与管理,2008,33(9):45-48.

[73]WU B,LIU Y,ZHEN J,et al. Protective effect of methionine on the intestinal oxidative stress and microbiota change induced by nickel[J]. Ecotoxicology environment safety,2022(244):1-10.

[74]WU B,CUI H,PENG X,et al. Toxicological effects of dietary nickel chloride on intestinal microbiota[J]. Ecotoxicology environment safety,2014(109):70-76.

[75]ZHOU X,LI J,SUN J L. Oral nickel changes of intestinal microflora in mice [J]. Current microbiology,2019,76(5):590-596.

[76]魏艳红,郭建强,陈志明,等. 环境汞污染对人体健康的影响及预防措施[J]. 大众科技,2014,16(3):59-61.

[77]ZHAO Y,ZHOU C,WU C,et al. Subchronic oral mercury caused intestinal injury and changed gut microbiota in mice[J]. Science of the total environment,2020 (721):1-10.

[78]TIAN X,LIN X,ZHAO J,et al. Gut as the target tissue of mercury and the extraintestinal effects[J]. Toxicology,2022(484):1-8.

[79]PINTO D V,RAPOSO R S,MATOS G A,et al. Methylmercury interactions with gut microbiota and potential modulation of neurogenic niches in the brain[J]. Frontiers in neuroscience,2020(14):1-5.

[80] ROTHENBERG S E,KEISER S,AJAMI N J,et al. The role of gut microbiota in fetal methylmercury exposure: insights from a pilot study[J]. Toxicology letters, 2016(242):60-67.

[81]ALHARBI O M L,BASHEER A A,KHATTAB R A,et al. Health and envi-

ronmental effects of persistent organic pollutants[J]. Journal of molecular liquids, 2018(263):442-453.

[82]戴惠玲.持久性有机污染物及其对人体健康的危害[J].中国医药导报,2008,5(17):101-102.

[83]李坤陶,李文增.持久性有机污染物对人体健康的危害[J].生物学教学,2006,(12):9-10.

[84] TOTTENBORG S S, HOUGAARD K S, DEEN L, et al. Prenatal exposure to airborne polychlorinated biphenyl congeners and male reproductive health [J]. Human reproduction update, 2022, 37(7): 1594-1608.

[85]WAHLANG B. Exposure to persistent organic pollutants: impact on women's health[J]. Review on environment health, 2018, 33(4): 331-348.

[86]KIM Y A, PARK J B, WOO M S, et al. Persistent organic pollutant-mediated insulin resistance[J]. International jounral environment research public health, 2019, 16 (3):448-448.

[87]TIAN Y, GUI W, RIMAL B, et al. Metabolic impact of persistent organic pollutants on gut microbiota[J]. Gut microbes, 2020, 12(1): 1-16.

[88]CHENG S L, LI X, LEHMLER H J, et al. Gut microbiota modulates interactions between polychlorinated biphenyls and bile acid homeostasis[J]. Toxicology science, 2018, 166(2): 269-287.

[89]RUDE K M, KEOGH C E, GAREAU M G. The role of the gut microbiome in mediating neurotoxic outcomes to PCB exposure[J]. Neurotoxicology, 2019(75): 30-40.

[90] PETRIELLO M C, HOFFMAN J B, VSEVOLOZHSKAYA O, et al. Dioxin-like PCB 126 increases intestinal inflammation and disrupts gut microbiota and metabolic homeostasis[J]. Environmental pollution, 2018(242): 1022-1032.

[91]POPLI S, BADGUJAR P C, AGWARWAL T, et al. Persistent organic pollutants in foods, their interplay with gut microbiota and resultant toxicity[J]. Science of the total environment, 2022(832): 1-14.

[92]SU H, LIU J, WU G, et al. Homeostasis of gut microbiota protects against polychlorinated biphenyl 126-induced metabolic dysfunction in liver of mice[J]. Science of the total environment, 2020(720): 1-16.

[93]KIM K H,JAHAN S A,KABIR E,et al. A review of airborne polycyclic aromatic hydrocarbons (PAHs) and their human health effects[J]. Environment international,2013(60):71-80.

[94]ZHANG W,SUN Z,ZHANG Q,et al. Preliminary evidence for an influence of exposure to polycyclic aromatic hydrocarbons on the composition of the gut microbiota and neurodevelopment in three-year-old healthy children[J]. BMC pediatrics,2021,21(1):1-13.

[95]REDFERN L K,JAYASUNDARA N,SINGKETON D R,et al. The role of gut microbial community and metabolomic shifts in adaptive resistance of Atlantic killifish (Fundulus heteroclitus) to polycyclic aromatic hydrocarbons[J]. Science of the total environment,2021(776):1-9.

[96]王洪雪,李光植.有机农药对人类和环境的危害[J].中外企业家,2019,(9):132-133.

[97]GIAMBO F,TEODORO M,COSTA C,et al. Toxicology and microbiota: how do pesticides influence gut microbiota? a review[J]. International journal environment research public health,2021,18(11):5510-5510.

[98]LI M,LIU T,YANG T,et al. Gut microbiota dysbiosis involves in host non-alcoholic fatty liver disease upon pyrethroid pesticide exposure[J]. Environment science ecotechnology,2022(11):1-10.

[99]NASUTI C,COMAN M M,OLEK R A,et al. Changes on fecal microbiota in rats exposed to permethrin during postnatal development[J]. Environmental science and pollution research,2016,23(11):10930-10937.

[100] LUO M,ZHOU D D,SHANG A,et al. Influences of food contaminants and additives on gut microbiota as well as protective effects of dietary bioactive compounds [J]. Trends in food science & technology,2021(113):180-192.

[101]ZHANG Q Q,YING G G,PAN C G,et al. Comprehensive evaluation of antibiotics emission and fate in the river basins of China: source analysis, multimedia modeling, and linkage to bacterial resistance[J]. Environmental science & technology, 2015,49(11):6772-6782.

[102]BEDFORD M. Removal of antibiotic growth promoters from poultry diets: implications and strategies to minimise subsequent problems[J]. World's poultry sci-

ence journal, 2000, 56(4): 347–365.

[103] CHEN J, YING G G, DENG W J. Antibiotic residues in food: extraction, analysis, and human health concerns[J]. Journal agriculture food chemistry. 2019, 67(27): 7569–7586.

[104] MARTINS V V, ZANETTI M O, PITONDO-SILVA A, et al. Aquatic environments polluted with antibiotics and heavy metals: a human health hazard[J]. Environment science pollution research international, 2014, 21(9): 5873–5878.

[105] HASAN T H, Al-HARMOOSH R A. Mechanisms of antibiotics resistance in bacteria[J]. System review pharma, 2020, 11(6): 817–823.

[106] NI H, CHAN K-WB, CHENG Q, et al. A novel clinical therapy to combat infections caused by hypervirulent carbapenem-resistant klebsiella pneumoniae[J]. Journal of infection, 2022, 85(2): 174–211.

[107] ZHENG S, QIU X, CHEN B, et al. Antibiotics pollution in Jiulong River estuary: source, distribution and bacterial resistance[J]. Chemosphere, 2011, 84(11): 1677–1685.

[108] NG K M, FERREYRA J A, HIGGINBOTTOM S K, et al. Microbiota-liberated host sugars facilitate post-antibiotic expansion of enteric pathogens[J]. Nature, 2013, 502(7469): 96–99.

[109] JAKOBSSON H E, JERNBERG C, ANDERSSON A F, et al. Short-term antibiotic treatment has differing long-term impacts on the human throat and gut microbiome[J]. PLoS one, 2010, 5(3): 1–12.

[110] DETHLEFSEN L, HUSE S, SOGIN M L, et al. The pervasive effects of an antibiotic on the human gut microbiota, as revealed by deep 16S rRNA sequencing[J]. PLoS biology, 2008, 6(11): 1–18.

[111] ARANDA-DIAZ A, NG K M, THOMSE T, et al. Establishment and characterization of stable, diverse, fecal-derived in vitro microbial communities that model the intestinal microbiota[J]. Cell host microbe, 2022, 30(2): 260–272.

[112] RAMIREZ J, GUARNER F, BUSTOS FERNANDEZ L, et al. Antibiotics as major disruptors of gut microbiota[J]. Frontiers cell infection microbiology, 2020(10): 1–10.

[113] PALLEJA A, MIKKELSEN K H, FORSLUND S K, et al. Recovery of gut

microbiota of healthy adults following antibiotic exposure [J]. Nature microbiology, 2018, 3(11): 1255–1265.

[114] VETHAAK A D, LEGLER J. Microplastics and human health [J]. Science, 2021, 371(6530): 672–674.

[115] YONG C Q Y, VALIYAVEETTIL S, TANG B L. Toxicity of microplastics and nanoplastics in mammalian systems [J]. International journal environment research public health, 2020, 17(5): 1–24.

[116] SCHWABL P, KOPPEL S, KONIGSHOFER P, et al. Detection of various microplastics in human stool: a prospective case series [J]. Annals of internal medicine, 2019, 171(7): 453–457.

[117] PRATA J C, DA COSTA J P, LOPES I, et al. Environmental exposure to microplastics: an overview on possible human health effects [J]. Science of the total environment, 2020(702): 1–10.

[118] YAN Z, LIU Y, ZHANG T, et al. Analysis of microplastics in human feces reveals a correlation between fecal microplastics and inflammatory bowel disease status [J]. Environment science technology, 2022, 56(1): 414–421.

[119] AMATO-LOURENCO L F, CARVALHO-OLUVEIRA R, JUNIOR G R, et al. Presence of airborne microplastics in human lung tissue [J]. Journal hazard materials, 2021(416): 1–6.

[120] LU L, LUO T, ZHAO Y, et al. Interaction between microplastics and microorganism as well as gut microbiota: a consideration on environmental animal and human health [J]. Science of the total environment, 2019(667): 94–100.

[121] LU L, WAN Z, LUO T, et al. Polystyrene microplastics induce gut microbiota dysbiosis and hepatic lipid metabolism disorder in mice [J]. Science of the total environment, 2018(631/632): 449–458.

[122] LI B, DING Y, CHENG X, et al. Polyethylene microplastics affect the distribution of gut microbiota and inflammation development in mice [J]. Chemosphere, 2020 (244): 1–10.

[123] TAMARGO A, MOLINERO N, REINSA J J, et al. PET microplastics affect human gut microbiota communities during simulated gastrointestinal digestion, first evidence of plausible polymer biodegradation during human digestion [J]. Scientific re-

ports, 2022, 12(1): 1-15.

[124] WEN S, ZHAO Y, LIU S, et al. Microplastics-perturbed gut microbiota triggered the testicular disorder in male mice: via fecal microbiota transplantation[J]. Environment pollution, 2022(309): 1-12.

第3章
肠道修复的利器——益生菌

2021年9月29日下午，中共中央政治局就加强我国生物安全建设进行第三十三次集体学习。中共中央总书记习近平在主持学习时强调："要促进生物技术健康发展，在尊重科学、严格监管、依法依规、确保安全的前提下，有序推进生物育种、生物制药等领域产业化应用。"

3.1 益生菌概况

3.1.1 益生菌的研究现状

益生菌是一类能够维持肠道微生态平衡，对人体各个器官和功能均有益处的细菌。益生菌代谢产生的物质称为益生元，也有益于人体。在人体内，益生菌广泛分布且数量众多，因此被称为"第二大脑"或"第二心脏"。在微生物研究日益深入的背景下，益生菌领域也得到了相应的发展，益生菌的国内外相关研究逐年增多。根据Web of Science数据库的数据统计，自2000年以来，与益生菌有关的论文发表数量稳步上升，益生菌研究成为当今世界最流行的研究领域之一。2020年，有关益生菌的研究文章数量已超过5500篇，而2021年则继续保持了较高的增长趋势，发表的论文总数已经超过6000篇。然而，2022年受新冠感染的影响，发文量有所降低（图3-1）。

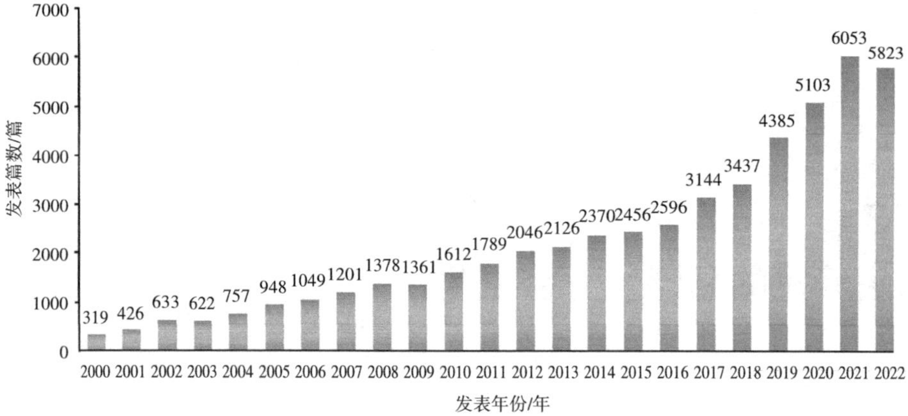

图3-1 自2000年起益生菌领域总发文量

就全球而言，欧美国家中美国、意大利、西班牙等国在益生菌研究方面较为深入，并且论文发表数量最多，亚洲主要是中国、印度、伊朗和韩国等国在该领域卓有建树。此外，南美洲的巴西也是益生菌研究领域的佼佼者。尽管中国在该领域起步较晚，但其发展势头已经位居世界首位，2022年的发文总量已超过美国，成为发表论文数量最多的国家。前10位发文总量排名中有5个亚洲国家，且其发文量的上升趋势均超过欧美国家。以上趋势充分体现了亚洲地区对于益生菌研究的浓厚兴趣（图3-2）。

目前，美国是益生菌领域发表高被引论文最多的国家，共发表231篇高被引文章，远超其余各国（图3-3）。然而，从每篇文章的引用频次看，爱尔兰以每篇440.81次位居榜首，而第2～8位均为欧美国家，中国仅位居第10位（图3-4）。这表明中国在益生菌领域发表的论文无论在影响力还是质量上都与世界先进水平存在相当的差距。

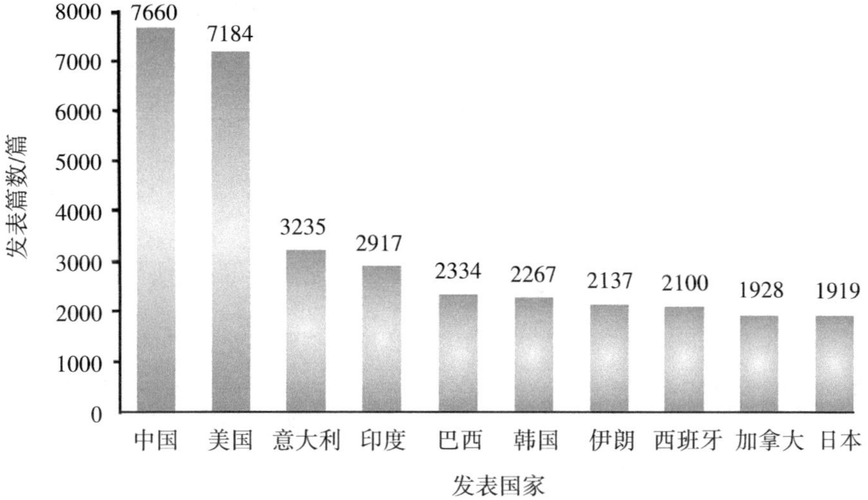

图3-2 益生菌领域发文排名前10位的国家

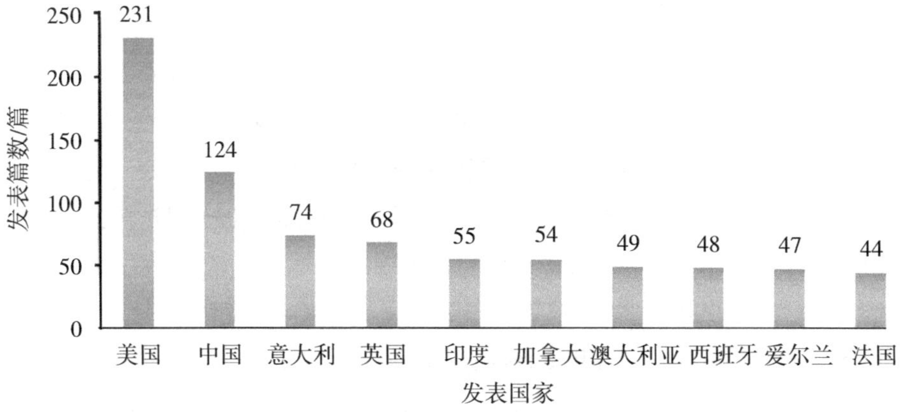

图3-3 益生菌领域发表高被引文章排名前10位的国家

尽管中国对益生菌的研究起步较晚，与欧美国家在研究积累、质量等方面存在差距，但最近几年，中国在益生菌领域的研究取得了突破性进展，高水平文章接连发表，越来越多的优质研究课题立项，在国际影响力和科研产出等方面，中国均处于领先地位。截至2022年1月30日，2021年中国发表的益生菌论文中共有13篇文章入选高被引论文。自2020年开始，中国的高水平论文数量开始大幅增加，且保持了这一增长势头，2021年高水平论文发表量已达78篇，相较于榜首美国的86篇文章，中国与世界前列的差距已经所剩无几了。

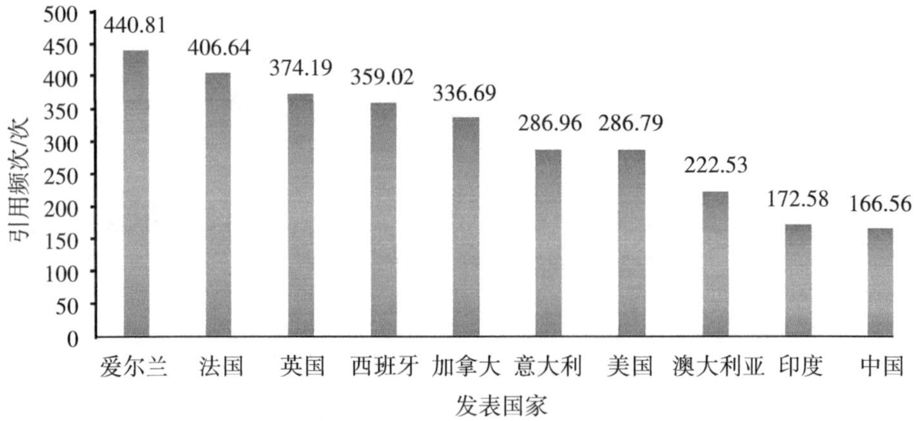

图3-4 益生菌领域论文年均被引频次排名前10位的国家

以上数据表明，在经济快速增长、科研投入日益加大的大环境下，中国的益生菌领域得到了蓬勃的发展。同时，世界前沿国家的益生菌研究已经进入平稳期，相信不久的将来，中国不仅在发文量上会超过美国，而且在品质方面也将成为世界领头羊。

3.1.2 热门益生菌

根据WOS论文类型分类结果，剔除荟萃分析、临床试验设计等文章，对2021年发表的165项临床研究使用的菌种和菌株进行统计[1]，结果显示在菌种水平上，鼠李糖乳杆菌（*Lactobacilus rhamnosus*）、嗜酸乳杆菌（*Lactobacilus acidophilus*）、植物乳杆菌（*Lactobacilus plantarum*）、长双歧杆菌（*Bifidobacterium longum*）、动物双歧杆菌（*Bifidobacterium animalis*）及嗜热链球菌（*Streptococus thermophilus*）是最受欢迎的菌种。在菌株水平上，前3个研究热点是鼠李糖乳杆菌GG、罗伊氏乳杆菌BB-12

(*Lactobacillus reuteri* BB-12)和乳酸双歧杆菌DSM 17938(*Bifidobacterium lactis* DSM 17938)。本章从不同角度出发，总结国内外学者关于以上热门益生菌作用及机制方面的最新研究进展。2021年的临床研究，主要探讨了鼠李糖乳杆菌GG对急性胃肠炎[1, 2]、糖尿病[3, 4]、肺炎[5]等的改善作用，乳双歧杆菌BB-12治疗代谢综合征[3, 6]、改善老年人身体状况的作用[7]，以及罗伊氏乳杆菌DSM17938对提高婴儿健康的作用[8-10]。还有动物双歧杆菌的乳酸亚种Bl-04[11, 12]、B420[13, 14]，植物乳杆菌PS128[15, 16]，鼠李糖乳杆菌HN001[17]和其他菌株也被广泛研究[18]。

3.2 益生菌的作用与机制

3.2.1 屏障作用

肠道屏障功能是指胃肠道上皮具有的防止致病性抗原侵入肠腔的分隔作用。肠道屏障能够有效地阻挡浓度高达1012个/g、超过500种肠道寄生菌及其毒素进入肠腔外组织和器官，从而保护机体免受内源性微生物及其毒素的侵害。维持正常肠道屏障功能需要肠黏膜上皮屏障、肠道免疫系统、肠道内正常菌群、肠道内分泌及蠕动的协同作用。肠道屏障防御系统包括IgA的分泌、抗菌肽的产生、黏膜层以及与上皮相连接的黏附复合物[19]。其中上皮细胞在屏障作用中处于中心位置，这些细胞从肠道管腔中接收分子信号，并且和基础免疫系统细胞进行交换，近年来发现一些非特异性免疫细胞也参与了该过程。肠道屏障防御系统是机体抵御外来病原体入侵及自身免疫反应的第一道防线，在肠易激综合征、炎症性肠病、感染性小肠结肠炎+腹腔疾病等多种胃肠道相关疾病的发病机制中起着重要的作用[20]。

鼠李糖乳杆菌GG与益生菌复合物VSL#3在小鼠及Caco-2肠细胞中的作用显示，鼠李糖乳杆菌GG会对肠上皮细胞产生一定的影响，从而维持肠道屏障功能。同时，鼠李糖乳杆菌GG还能刺激肠道中相关免疫应答反应，增强肠腔内微生物菌群多样性，促进细菌增殖等。鼠李糖乳杆菌GG长期存在于胃肠道中，与体内黏液结合域菌毛的表达相关[21]。肠道微生态失调导致了多种疾病如肥胖、糖尿病及癌症等。体外研究发现，鼠李糖乳杆菌GG及其可溶性因子（p75和p40）可通过激活Akt和抑制核因子κB（nuclear factor kappa-B，NF-κB）而抑制上皮细胞凋亡，从而保护肠道屏障[22]（图3-5）。

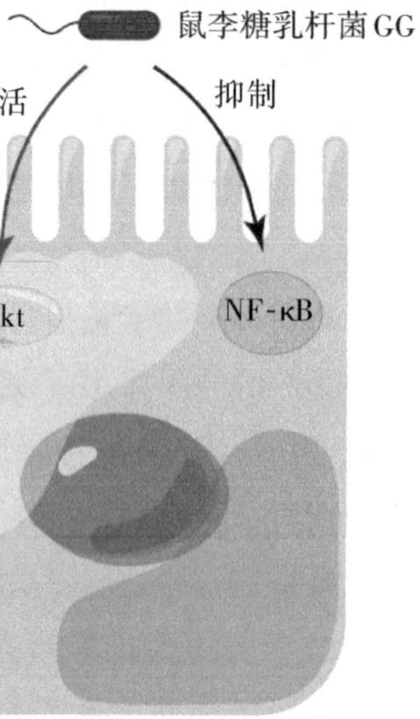

图3-5　鼠李糖乳杆菌GG影响肠道屏障的机制

植物乳杆菌、干酪乳杆菌、鼠李糖乳杆菌、嗜酸乳杆菌等能够通过刺激人黏膜基因对人体防御网络中的不同途径进行调控，从而增强人体的防御能力。研究表明，微生物通过影响宿主防御系统来维持机体正常生理活动，其中就包括了一个NF-κB信号级联激活因子上调，该信号因子涉及B细胞成熟及淋巴生成有关基因转录。研究人员将乳酸菌、双歧杆菌、链球菌作为益生菌对旋毛虫感染引起的肠易激综合征进行治疗，发现在小鼠模型中NF-κB信号因子显示出阳性结果[23]。对双歧杆菌或乳酸杆菌处理肠易激综合征小鼠的腹部收缩反应和退缩反射评分，发现d-乳酸水平降低，血浆二胺氧化酶（diamine oxidase，DAO）

浓度降低，症状得以缓减。另外，有研究表明，使用益生菌后，促炎细胞因子白细胞介素-17（interleukin-17，IL-17）和白细胞介素-6（interleukin-6，IL-6）被抑制，细胞紧密连接的闭塞蛋白和黏连蛋白的表达被增强了[24]。

3.2.2 黏附作用

益生菌与肠黏膜的黏附是益生菌的重要特性，也是益生菌在肠内定植、拮抗病原体和免疫系统变异必需的特性。各种乳酸菌蛋白结合糖基或磷脂能够改善自身的黏附力，提高对黏液细胞的黏附[25，26]。从罗伊氏乳杆菌中提取的细菌黏附蛋白——黏液结合蛋白（mucus-binding protein，MUB）的研究表明[27]，植物乳杆菌等益生菌可以通过诱导黏蛋白MUC2（mucin2，MUC2)和MUC3（mucin3，MUC3)防止肠致病性大肠杆菌的附着。此外，由于益生菌在肠道上皮表面附着，黏附位点被阻断，从而使病原体难以在肠道上定植[28]。研究表明，在摄入乳酸菌后，它会与病原菌共同竞争结合位点，从而保护肠道不被感染[29]（图3-6）。

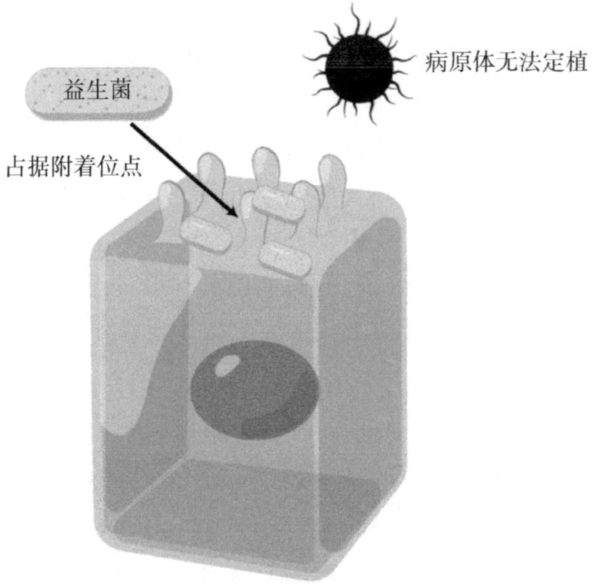

图3-6 益生菌通过黏附作用保护肠细胞

由于益生菌双歧杆菌具有耐酸性，可以增强肠道稳定性，并改善肠道表面性能。与对酸敏感的菌株相比，来自长双歧杆菌和链状双歧杆菌的耐酸菌株对人体肠道黏液具有有效的附着作用[30]。益生菌与VSL#3联合使用可以促进黏蛋白的合成，同时促进黏蛋白基因的表达，从而增强细菌对肠上皮的附着。在未分化和分化的角化细胞中，罗伊氏乳杆菌ATCC 55730和鼠李糖乳杆菌AC413能够显著降低金黄色葡萄球菌感染引起的角化细胞死亡数量，因此，可以利用上述益生菌预防金黄色葡萄球菌感染和抑制其增殖[31]。

3.2.3 养分竞争

养分竞争可能是抑制病原体在人体肠道中定植的机制之一。这种竞争有两种不同的方式：第一，防止肠道内生长增殖所需的营养和能量被病原体吸收；第二，通过发酵和代谢产生的代谢产物，如短链脂肪酸和有机酸，降低肠道的pH值，使大多数病原体如大肠杆菌和沙门氏菌生长受到限制[25]。由于双歧杆菌S2-1在维生素K和其他生长因子上的养分竞争胜过牙龈卟啉单胞菌（*Porphyromonas gingivalis*），故双歧杆菌S2-1可阻止牙龈卟啉单胞菌生长[32]。在无菌小鼠（定植人类婴儿菌群）中定植益生菌如副干酪乳杆菌（*Lactobacillus paracasei*）或鼠李糖乳杆菌后，观察到其短链脂肪酸、氨基酸和甲胺代谢等通路发生明显变化[33]。

另外，益生菌德氏乳杆菌（*Lactobacillus delbrueckii*）和嗜酸乳杆菌还可以将氢氧化铁结合到细胞表面，从而避免了病原体对氢氧化铁的利用[34]。而鼠李糖乳杆菌和副干酪乳杆菌这两种益生菌通过竞争、生态位互换和排斥等不同机制抑制致病性单核增生李斯特菌（*Listeria monocytogenes*）和鼠伤寒沙门氏菌（*Salmonella typhimurium*）等致病菌的生物膜形成过程[35]。研究表明，在上述益生菌的影响下，致病性单核增生李斯特菌的生物膜细胞减少超过3个对数周期。

3.2.4 抑制性小分子

益生菌通过产生各种抗菌物质对革兰氏阴性菌和革兰氏阳性菌起到抗菌作用，这些物质包括有机酸[36]、细菌素[37-39]、二乙酰[40]、乙醇[41]、过氧化氢[36]和二氧化碳[42]等。益生菌可产生的抑制性不分子如表3-1所示。

表3-1 益生菌可产生的抑制性小分子[43]

抑制性小分子	范例	所属菌株
细菌素	Pediocin PA-1	*P. acidilactici*
	Nisin	*L. lactis subsp. lactis*
	Enterocin AS48	*E. Faecalis*
	肠溶素	*E. Faecalis*
细菌素类抑制物质	—	*L. lactis subsp. lactis* CECT-4434
	—	*P. acidilactici* Kp10
	—	*L. mesenteroides* 406
抗生素	Reuterin	*L. reuteri* DSM 20016
	红霉素	*L. reuteri*
有机酸	乳酸、乙酸	LAB
过氧化氢	—	*P. acidilacti*, *L. mesenteroides*, *L. brevis*, *L. brevis*, *L. casei*
其他	乙醇	*Bifidobacterium longum*
	二乙酰	*L. plantarum*, *L. helveticus*, *L. bulgaricus*, *E. Faecalis*, *L. mesenteroides*
	二氧化碳	*Heterofermentative*, LAB

细菌素抑制病原菌生长的主要机制有两种：一是抑制其细胞壁生长，二是防止其形成膜中孔。Nisin是一种与细胞壁前体和脂质Ⅱ复合物形成有关的抗菌化合物，可以抑制细胞壁的合成，也可以通过去除聚集的复合物

和结合肽防止膜中孔的形成[37]。此外，细菌素能促进肠上皮细胞增殖并降低其对抗生素的敏感性，对胃肠道病原体有抑制作用[43]。许多细菌素还能诱导蛋白酶活性或分泌抗菌药，如乳酸脱氢酶抑制剂和β-内酰胺类抗生素等[44，45]。嗜酸乳杆菌能产生酸德林等多种抗菌化合物，这些抗菌化合物能够抑制肠道致病菌的生长并阻止其增殖，从而使宿主免受肠内毒素引起的损害。嗜酸西林、乳酸菌素与植物乳杆菌又能产生抗菌化合物——乳酸菌素[46]。唾液乳杆菌UCC118株产生的细菌素Abp118可激发抗菌反应，对单核增生李斯特菌所感染的小鼠具有保护作用[47]。研究表明，乳酸菌和双歧杆菌对幽门螺杆菌、大肠杆菌、单核增生李斯特菌、轮状病毒和沙门氏菌具有抑制作用[23]。双歧杆菌的几个菌株（双歧杆菌NCFB 1454）能够产生出一种独特的细菌素（双歧杆菌素B），该细菌素能够抑制革兰氏阳性菌生长，尤其对大肠杆菌C1845和沙门氏菌具有较高的抑制性[43]。

3.2.5 毒性降低机理

益生菌降低毒性的机理主要包括：对毒素受体进行酶促修饰，使宿主减少对毒素的吸收，同时减少毒素产生、降低肠道pH值及毒性等，还可以进一步预防该毒素引起的疾病[23]。博拉酵母菌具有降解家兔回肠艰难梭菌毒素受体的能力，并可通过产生多胺防止大鼠空肠的霍乱引起的分泌。Trick等[48]在C57BL/6小鼠中研究了一种共生菌制剂（*L. casei* F8、*L. plantarum* F44、*B. lactis* 8：8、*B. breve* 46、抗性淀粉、低聚异麦芽糖和低聚半乳糖）对艰难梭菌感染（clostridium difficile infection，CDI）的影响。配方喂养后，乳酸菌和双歧杆菌数量增加，但未检测到任何盲肠毒素。盲肠qPCR检测后发现艰难梭菌DNA拷贝明显减少。

3.2.6 调节肠道菌群

人类肠道微生物时刻在对多种生物功能进行调控。我们对其的认识逐步从能量调节转向增强宿主对有害微生物的免疫力与中和毒素的能力。随着人们对其生物学特性认识的加深，越来越多的研究开始关注如何通过调控肠道微生态影响人体健康。益生菌和益生元可能的用途一直都是与保持宿主理想肠道健康有关，治疗/防止宿主反复炎症及免疫系统相关疾病等[49]。益生菌被广泛应用于防治若干导致或关联肠道微生物群失调疾病，例如，急性感染性腹泻、抗生素相关性腹泻（antibiotic-associated diarrhea，AAD）等，以及其他胃肠道疾病，例如绞痛或者肠易激综合征等。益生菌对肠内细菌群及免疫应答具有保护作用，用益生菌处理后肠道微生物群落构成变得稳定，这表明益生菌与疾病康复呈正相关[50]。

3.3 益生菌的应用现状

3.3.1 益生菌改善人体代谢异常

（1）益生菌可调控胆固醇水平

益生菌会影响血液中的脂质水平。肠球菌属、拟杆菌属和链球菌属可以通过产生短链脂肪酸[51, 52]来调节宿主胆固醇的合成[53, 54]、吸收与排放[55]，从而调控宿主胆固醇水平。乳酸菌能够产生胆盐水解酶[56]，丰富肠道胆汁酸池，抑制胆汁酸盐的重吸收，从而减少胆固醇的合成并使游离的胆汁酸通过粪便排出[57]。此外，益生菌促前列腺寡聚真杆菌（*Eubacterium coprostanoligenes*）还可直接降解肠道胆固醇[58]。

（2）益生菌可改善肥胖相关病症

D. welbionis J115′是2021年发现的一种新的瘤胃球菌，其在改善肥胖及相关代谢性疾病方面具有显著的作用[59]。*D. welbionis* J115′广泛存在于普通人群中，在患有代谢综合征的肥胖人群中，该菌的相对丰度与体重指数和空腹血糖值呈负相关。研究表明，向高脂饲料喂养的小鼠灌胃 *D. welbionie* J115′菌株，能够显著减少小鼠体重和脂肪增加量，改善糖耐量，降低胰岛素抵抗。研究还

发现该菌能增加小鼠的线粒体数量，抑制高脂饮食诱导的棕色脂肪组织白化，从而帮助机体降低血糖水平和体重，除此之外，该菌也具有抗炎效应。这些效果对于2型糖尿病患者及肥胖患者而言都是非常重要的。不过总的来说，*D. welbionis* J115′或许可直接且有益地影响机体代谢，有望成为针对肥胖及相关代谢性疾病的下一代益生菌的强有力候选者（图3-7）。

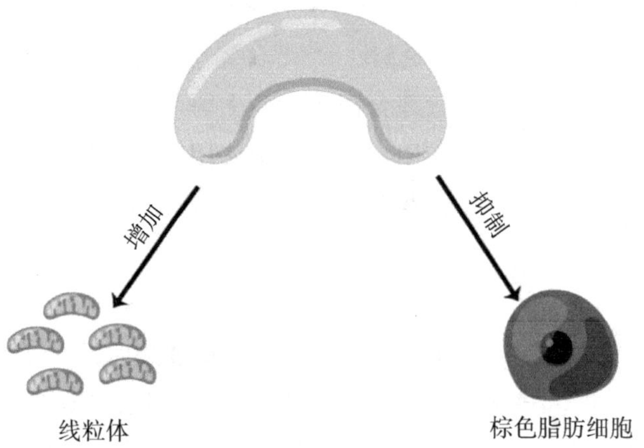

图3-7　*D. welbionis* J115'治疗肥胖相关疾病的机制

（3）益生菌可协助降低高血压

干酪乳杆菌和植物乳杆菌菌株具有潜在的降压作用。对大鼠的研究报告表明，在使用益生菌后，大鼠血压显著降低。进一步研究发现，乳蛋白等蛋白质降解后，生物活性化合物的释放具有降压作用，这一作用会影响肾素-血管紧张素系统[60]。因此，益生菌的蛋白质降解能力可以通过刺激血管紧张素转换酶（angiotensin converting enzyme，ACE）抑制肽的分泌来帮助降低血压。

3.3.2 益生菌治疗胃肠道疾病

（1）抗生素相关性腹泻

抗生素相关性腹泻是指使用抗生素后引起肠道菌群失调而导致的常见的医源性腹泻。抗生素相关性腹泻的发病率为5%～30%。老年人群的发病率更高，尤其是住院老年患者，抗生素相关性腹泻也见于社区和养老机构的老年患者。高危患病人群抗生素相关性腹泻的发生率可高达9.2%～36.79%。抗生素相关性腹泻可以被几乎所有的抗生素诱发[61，62]，这主要是由于抗生素的副作用所致，因为使用抗生素会导致胃肠道环境改变、稳定性下降，同时肠道菌群的多样性和丰度也会降低[63]。这些剧烈的变化导致了正常肠道细菌的消亡，并提高了艰难梭状芽孢杆菌等病原体的定植机会。

一项关于在老年患者（65岁以上）和成人（18～64岁）中使用益生菌治疗抗生素相关性腹泻的系统综述研究通过评估30项符合标准的随机管理测试，对益生菌治疗抗生素相关性腹泻的效果做了系统性的评估。临床研究表明，益生菌作为抗体的佐剂，可降低成人中抗生素相关性腹泻的发生概率，但对老年人的效果有所减弱[64]。普洛斯特罗的研究证明，一些益生菌菌株如布拉氏链球菌和鼠李糖乳杆菌GG参与了预防抗生素相关性腹泻，但其他菌株如保加利亚乳杆菌、德氏乳杆菌和唾液链球菌不能预防抗生素相关性腹泻[65-67]。

益生菌在许多临床试验中被证明可以预防/治疗抗生素相关性腹泻，其中的分子机理也有相应的报道[68，69]。益生菌通过调节肠道菌群、改变营养物质和胆汁酸代谢、调控胃肠道转运活动、支持肠道屏障功能和影响免疫系统等方式降低抗生素相关性腹泻的患病风险。抗生素治疗过程中，会对胃肠道菌群组成的生物屏障造成破坏，引起胃肠道环境失衡，从而导致机会性病原体能够更容易定植并快速生长。抗生素治疗还会降低胃肠消化功能，导致未消化的碳水化合物积累，并使体内代谢产生紊乱，短链脂肪酸和修饰性胆汁酸的水平降低。益生菌可通过直接抑制病原体的生长，或通过促使短链脂肪酸合成、产生其他分泌代谢物（如细菌素）以及降低肠腔pH值和氧气水平诱导肠道微生物群组成的改变，以对抗胃肠道中抗生素产生的不利影响[70]。益生菌还可引起胆汁酸成分的变化，并直接与肠上皮和免疫系统相互作用，从而增强肠道屏障功能，更好地调节水和溶质的转运[71]（图3-8）。

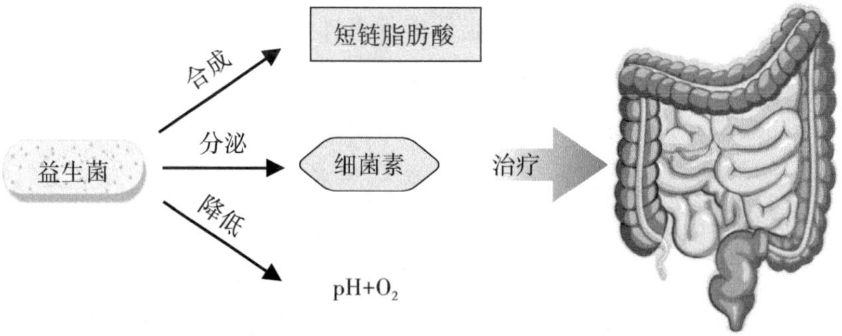

图3-8 益生菌治疗抗生素相关性腹泻的机制

（2）肠易激综合征

最近发表的一些生理学、流行病学和临床研究数据表明，肠道微生物群参与了肠易激综合征的发病机制，然而，肠易激综合征的病理生理机理仍未被发现[72, 73]。有研究表明，改变宿主与益生菌结合的宿主肠道微生物可以影响宿主的一些肠道功能，如敏感性和运动性，这可能影响肠易激综合征的发病机制[74]。一项临床研究结果表明，与安慰剂组相比，摄入长双歧杆菌婴儿亚种的患者组疾病症状显著改善，同时血清IL-10/IL-12恢复正常，表明益生菌可以帮助缓解与肠易激综合征相关的炎症状态[75, 76]。此外，一些植物乳杆菌在缓解肠易激综合征患者的少数症状方面也有显著的作用，其中，DSM 9843菌株显著降低了肠胃胀气，299 V和LPO1菌株明显降低了肠道疼痛[77-79]。

（3）溃疡性结肠炎

临床试验表明，使用鼠李糖乳杆菌GG治疗溃疡性结肠炎可能比标准治疗会更有效地预防疾病的复发[80]。大肠杆菌Nissle 1917在避免成人溃疡性结肠炎复发方面表现出与抗生素5-氨基水杨酸相似的效果[81]。

（4）贮袋炎

贮袋炎是指急性和慢性溃疡性结肠炎患者在经历了恢复性直肠结肠切除术后，创口恢复过程中的回肠袋表现出的炎症。一些使用益生菌的临床试验表明，益生菌在减轻贮袋炎症方面具有很高的安全性和有效性，以及随后的抗生素治疗中，也有助于减少贮袋炎的复发和预防急性贮袋炎[82, 83]。科克伦合作组织的一项系统综述显示，与安慰剂相比，混合益生菌VSL#3在维持减轻慢性贮袋炎和避免贮袋炎的发作方面非常有效[84]。

3.3.3 益生菌辅助治疗新冠

新冠是新型冠状病毒感染的简称，是近年的一种急性感染性疾病。研究人员发现，鼠李糖乳杆菌GG、副干酪乳杆菌DG和LPC-S01具有免疫调节作用和抗新型冠状病毒活性[85]。研究表明，副干酪乳杆菌DG具备预防新型冠状病毒感染的潜力。副干酪乳杆菌DG可显著诱导参与抗病毒免疫的基因的表达，并阻止新型冠状病毒感染引发的促炎基因的表达，且在体外可显著抑制新型冠状病毒感染。Wu等[86]对比鉴定了新冠感染患者的肠道菌群及上呼吸道菌群的特异性变化，并分析了患者在接受益生菌辅助治疗后的肠道菌群恢复情况，结果发现，接受益生菌辅助治疗的新冠感染患者在治疗结束后粪便菌群组成得到恢复，主要表现为新冠感染相关菌群［大肠杆菌、肺炎克雷伯菌（*Klebsiella pneumoniae*）等］丰度的减少和健康肠道菌群［柔嫩梭菌（*Faecalibacterium*）等］的增加。Xu等[87]研究发现小鼠在接种新冠疫苗6个月后口服益生菌（植物乳杆菌GUANKE）可显著延长特异性抗体应答，并且在接种新冠疫苗后立即口服益生菌也可显著增强特异性体液免疫及细胞免疫应答。

3.3.4 益生菌辅助治疗肿瘤

肠道微生物在调控宿主免疫中起关键作用，紊乱的肠道菌群与宿主免疫相互作用诱发炎症，持续的炎症反应又加剧了肠道菌群的紊乱[88]。反过来，通过调控微生物，也可以对肿瘤治疗产生积极的影响[89]。因此，利用益生菌对肠道微生物调节，从而间接对肿瘤治疗提供有力的帮助，这可能是未来研究的方向。Si等[90]发现，鼠李糖乳杆菌GG与免疫检查点阻断药物抗PD-1联合使用后，癌症小鼠模型的抗肿瘤反应得以改善，而口服活的鼠李糖乳杆菌GG后，肿瘤浸润树突状细胞（dentritic cell，DC）和T细胞均显著增加，这表明小鼠抗肿瘤活性显著提高。Canale等[91]通过构建大肠杆菌Nissle 1917构建了工程益生菌，使其定植于肿瘤之上，将肿瘤组织产生的含氮废物转化为L-精氨酸，这使得抗PD-1疗效得以提升。

除免疫疗法外，益生菌在诸如化疗和靶向疗法等治疗中，无论是治疗过程，还是人体恢复过程，也都发挥了相当重要的功效[92]。研究表明益生菌可通过促进药效、清除损害抗癌效果的物质及降低毒性来调节宿主对化疗药物的响应[93]。吉西他滨是用于治疗癌症的常用药物之一，Panebianco等[93]的研究发现特定的益生菌给药可以恢复一些有利的菌群，从而缓解经吉西他滨治疗的胰腺癌小鼠产生的一些不良反应。玻玛西林是一款治疗乳腺癌的新药，80%～90%的患者使用此药后会产生腹泻。研究表明单独使用双歧杆菌，或将其与马来酸曲美布汀（一种调节胃肠道功能的药物）联合使用，能够显著缩短玻玛西林导致的腹泻的持续时间，并能预防更加严重的腹泻，同时并不会使患者产生便秘[92]。

3.3.5 益生菌治疗肾脏疾病

近年来，越来越多的研究表明，肠道菌群参与肾疾病的发生发展过程。“肠-肾轴”理论成为研究急性肾损伤和慢性肾病病理生理机制的热点。肠道菌群能通过调节机体内环境平衡促进细胞增殖分化及免疫应答等作用改善肾功能，并与许多肾病相关因素密切相关。但肠道菌群改变对肾病发生发展有何种影响尚需阐明。此外，益生菌可能对肾具有保护作用，但具体效果尚未明确。Zhu等[94]结合小鼠动物实验和临床试验，揭示了肠道菌群失调和益生菌干预在急性肾损伤和慢性肾病中的作用和机制，表明补充干酪乳杆菌Zhang能通过改善肠道菌群和增加有益代谢物（如短链脂肪酸和烟酰胺）来减轻肾损伤、减缓肾功能衰退。相关成果具有临床应用前景，为益生菌干预治疗肾病添加了新证。

3.3.6 益生菌治疗抑郁和焦虑

抑郁和焦虑是人类最常见的两种心理健康疾病，也是世界性难题。肠道和大脑通过一种叫做肠-脑轴通路的特定途径相互作用、相互影响，其中包括免疫系统、内分泌系统和神经系统。研究发现，采用乳酸杆菌和双歧杆菌混合治疗，能够有效缓解患者的焦虑症状[95]。但另一项研究结果显示，采用含长双歧杆菌BL-04、植物乳杆菌、发酵乳杆菌LF-16和鼠李糖乳杆菌LR-06的益生菌混合物对抑郁症患者进行治疗，对睡眠质量和抑郁情绪状态都没有提供任何积极的影响[96]。因此，需要更多重要的临床试验来探索益生菌对抑郁和焦虑的影响。

3.3.7 益生菌提高免疫力

受到病原体或外来物质入侵时，人体免疫系统通过先天性免疫和适应性免疫来抵抗这些有害物质的侵害[97]。益生菌能够直接影响巨噬细胞、T细胞（细胞介导的反应）、B细胞（体液反应）和树突状细胞，从而调节免疫应答。乳酸菌的代谢产物可调控Toll样受体（toll like receptor，TLR）、树突状细胞和自然杀伤细胞（natural killer cell，NK）等的免疫调节作用，同时也影响淋巴细胞增殖，平衡辅助性T细胞（Th1/Th2）的反应和特异性IgA的分泌等。枯草芽孢杆菌B10和博拉酵母菌针对特异性Toll样受体和相关因子有明显的调控能力，因此在控制鸡骨髓树突状细胞的免疫功能中起主要作用[98]。益生菌附着在树突状细胞表面后，可以观察到*MHC*-Ⅱ、*CD*40、*CD*80和*CD*86基因表达水平的上调。此外，可观察到*TLR*1、*TLR*2、*TLR*4和*TLR*15的表达增强，下游相关因子*TRAF*6、*MyD*88、*NF-κB mRNA*和*tab*1的水平升高。另一项体内研究报道了长双歧杆菌AH1206、唾液乳杆菌AH102和短双歧杆菌AH1205对诱导*Foxp*3+*treg*反应的菌株的特异性效应。据报道，长双歧杆菌AH1206可以通过增加Foxp3+treg的数量来改善婴儿、成人和无菌动物的气道炎症[46]。

肠道中健康微生物的积累和生长会促使多种免疫机制的成熟，特别是IgA和IgM分泌细胞循环的成熟。机体完成生理准备后，B细胞向效应位点移动，并积极地增殖，然后局部刺激各种细胞因子和分泌IgA。因此益生菌一旦进入肠道就会刺激IgA的产生。一项对无菌小鼠的研究证实了益生菌能够促进免疫系统中IgA的产生[99]。多项研究表明，由于双歧杆菌、乳杆菌等乳酸菌及其发酵产物的作用，先天免疫和适应性免疫在缓解过敏、预防胃黏膜病变发展、防御肠道病原体感染的同时得到提高[98]。

另外一项研究表明，通过喂给1.4岁的大鼠鼠李糖乳杆菌，导致其与*Th*1/*Th*2失衡相关的免疫衰老增强，对老年小鼠大肠杆菌感染的抵抗力增强，并且还发现喂给16个月大的小鼠鼠李糖乳杆菌后，小鼠抗氧化能力得到显著提高。此外，在益生菌喂养的小鼠中，干扰素-γ（interferon-γ，IFN-γ）水平升高，白细胞介素-4（interleukin-4, IL-4）和白细胞介素-10（interleukin-10，IL-10）

产生水平降低，吞噬和中性粒细胞呼吸爆发酶增加，但血浆单核细胞趋化蛋白-1（monocyte chemoattractant protein-1，MCP-1）和TNF-α水平未升高。在饲喂益生菌的小鼠中，免疫球蛋白E（immunoglobulin E，IgE）水平、免疫球蛋白G1（immunoglobulin G1，IgG1）和免疫球蛋白G2a（immunoglobulin G2a，IgG2a）的比值随着抗氧化酶活性的增加而降低，大肠杆菌向小鼠器官的易位也显著减少[100]。

3.4 益生菌市场发展面临的机遇与挑战

3.4.1 益生菌市场前景

全球益生菌市场正处于快速发展阶段，预计未来几年将继续保持良好的增长势头。益生菌市场的增长受到人们对健康和营养的日益关注以及消费者对天然、有机和功能性食品需求增加的推动。随着人们健康意识不断提升，越来越多消费者选择通过益生菌预防或治疗疾病，从而促进了益生菌在食品行业中的应用和发展，尤其是作为食品添加剂而出现。根据Grand View Research机构的数据（表3-2），2021年，全球益生菌的市场规模为581.7亿美元，预计2021—2030年，全球益生菌市场每年复合增长率将达到7.5%，2029年前，将超过千亿美元关口，到2030年将达到1115亿美元。这种增长是由5种驱动因素造成的：①健康饮食趋势。消费者对健康和营养的关注度提高，促使益生菌和益生菌制品的需求增加。②生活方式改变。生活压力、不良饮食、抗生素滥用等因素导致肠道菌群失衡，需要益生菌维护肠道健康。③全球老龄化趋势。随着人口老龄化，维护老年人的肠道健康变得更加重要，益生菌制品能够提供较好的肠道健康维护。④增强免疫力。益生菌可以增强免疫力，减少疾病风险，因此越来越多的人开始关注益生菌的作用。⑤市场竞争。随着市场的扩大和竞争的加剧，益生菌制品的价格不断下降，也促进了消费者对益生菌的需求。

表3-2 全球益生菌市场报告

报告内容	详细说明
2022年市场规模价值	619.5亿美元
至2030年规模预测	1112.1亿美元
年复合增长率	7.5%
主要评估国家	美国、加拿大、德国、墨西哥、中国、英国、法国、印度、日本、巴西、沙特阿拉伯

就区域市场而言，益生菌的竞争主要集中在亚洲、北美和欧洲三个大区域。随着中国进入经济新常态，人民生活水平不断提高，消费者对健康的重视程度也不断增强。根据中国食品科学技术学会益生菌分会的数据，亚太地区是全球消费规模最大的地区之一，其消费量占全球的44.4%，远高于欧洲（23.5%）和北美（17.8%），彰显了亚太地区益生菌行业的繁荣发展态势。就应用行业而言，世界益生菌下游应用产业主要可分为以乳制品为代表的功能性食品、营养保健品和药品，以及动植物用益生菌三大类。其中，功能性食品占比最高，达86%，营养保健品和药品的占比较低。

中国已成为全球第二大益生菌消费市场，预计在政策利好的情况下，将迎来千亿级市场。根据中国保健协会的统计数据，2020年中国益生菌产品的市场规模约为879.8亿元。根据欧睿国际的数据显示，中国益生菌的消费市场规模已超过意大利，居于全球第二位，并且以每年11%～12%的速度快速增长，预计将突破千亿大关。此外，随着消费者对食品安全意识的不断增强以及对健康食品需求的日益增长，益生菌市场呈现出快速扩张的趋势。中国益生菌市场将不断发展壮大，这是因为从科学基础上看，益生菌研究在中国已积极展开，为产业创新带来了持续动力；从政策层面看，2017年中国出台了《国民营养计划（2017—2030年）》，第一次建议将“营养纳入各项健康政策”，中共中央总书记习近平在党的十九大报告中正式提出“健康中国”战略，随后有关合理营养的各项政策被密集推出，它们相互衔接呼应，营造了益生菌产品发展的良好环境和机遇。消费者对健康的关注度越来越高，也为益生菌市场的发展提供了广阔的空间和良好的机遇。

3.4.2 消费者对益生菌产品的消费态度

消费者对益生菌产品的消费态度主要可以归纳为5点：①关注健康。随着人们健康意识的提高，消费者越来越注重饮食健康，对益生菌产品的需求也在增加，消费者普遍认为益生菌可以帮助维持肠道健康，提高免疫力等。②寻求高品质。随着益生菌市场的扩大和竞争的加剧，消费者越来越注重产品的品质，他们希望购买高品质、安全、可靠的益生菌产品。③喜欢方便易用的产品。消费者对于方便易用的益生菌产品有较高的偏好，如口服益生菌制剂、乳酸菌饮料等。④注重个性化需求。消费者在选择益生菌产品时，越来越注重个性化需求，例如需要特定的菌种，或者是对某些成分有过敏反应等。⑤注重口碑和推荐。消费者更倾向于购买口碑好、有良好推荐的益生菌产品，他们会通过朋友、亲戚的评价或者网上的评论等了解产品的真实情况，做出购买决策。

对健康和营养高度关注的人是益生菌产品的主要消费群体。益生菌可帮助消化不良人群改善肠道菌群平衡，缓解腹泻、便秘、胃炎等消化不良症状。对免疫系统较弱人群而言，益生菌可增强免疫系统的功能，提高人体抵抗力，预防疾病。在婴幼儿及孕产妇方面，益生菌可帮助婴幼儿建立健康的肠道菌群，预防肠道感染，益生菌对孕妇和哺乳期妇女也有益处。而对于长期服用抗生素的人群来说，益生菌可缓解抗生素对肠道菌群的破坏，降低患上抗生素相关腹泻的风险。对于身体健康的人群，益生菌可帮助维持肠道菌群平衡，促进身体健康，预防疾病。益生菌产品还受到了许多健身爱好者、养生人士、老年人等人群的青睐。总的来说，益生菌产品的受众面广，覆盖人群多样化。

3.4.3 益生菌领域未来展望与挑战

益生菌研究涉及广泛的研究领域，包括微生物学及其应用技术、食品科学与营养学、胃肠病学与肝病学、农业、乳品及畜牧科学，以及兽医学等。近10年与营养学、多学科科学相关的益生菌研究占比明显增加，而胃肠病学与肝病学虽然依然是益生菌研究主要涉及的学科，但是其占比有所下降。这一结果凸显出益生菌及其研究的两个重要特征：

（1）人体是自身和微生物组成的生态系统

营养不仅是人体与微生物组之间的纽带，也是两者相互作用的重要物质基础。人体与微生物组之间的关系取决于平衡，过度或不足都会影响健康。越来越多的研究显示，益生菌在营养方面扮演着重要角色[7-9]。在短期内，我们可以通过益生菌治疗或辅助治疗一些疾病，但在更长的时间范围内，我们更需要加强对益生菌营养属性的认识，利用该属性来治未病和提高人类身体素质，以践行“人类营养命运共同体”的理念。例如，我们可以将更多营养丰富的益生菌纳入配方中，开发更全面的特殊膳食或特殊医学用途配方食品。这将有助于保持人类、微生物和营养的动态平衡，从而实现“以菌养人”的目标。

（2）益生菌研究是一个涉及多学科的研究领域

益生菌研究是一个跨学科的研究领域，需要整合多个学科的知识和技术。为了探究微生态失衡与多种疾病之间的关系，益生菌研究需要结合相关疾病的病理学特征。同时，随着基础科学的不断发展，越来越多的新兴技术被应用到益生菌研究中，例如全基因组测序、单细胞测序、宏基因组学、转录组学、蛋白质组学和代谢组学等技术，可以深入揭示宿主的遗传信息、进化关系、生理特性、代谢网络、功能机制以及工业应用潜力。此外，培养组学和合成生物学（synthetic biology，SB）等技术为开发具有特定功能的新一代益生菌提供了方法学支持。新算法的开发可以整合多组学数据，从而为益生菌机制层面的研究助力。在产业化应用方面，益生菌的包埋、干燥、递送和检测等方面也需要新的技术支持。毫无疑问，只有在多个学科的支持下，益生菌领域才能继续蓬勃发展。

步入2000年以后，随着健康意识的不断增强，人们对于益生菌的认识也在逐渐提高，国内益生菌基础研究及产业化升级步伐加快。近年来，中国政府出台多项政策鼓励益生菌的推广与应用，消费市场对于益生菌的重视程度持续提升，增加了对益生菌产品的需求。在这种情况下益生菌行业迎来了又一轮淘汰和提升，原料及产品研发成了企业生存和持续发展至关重要的问题。

（3）益生菌的行业价值链

益生菌产业的价值链是一个包含4个主要环节的复杂系统。

首先是上游环节，也称为益生菌菌株的研究和开发环节。这个环节的重点是研究和评估益生菌菌株，找到有特殊功能和对人体有益的菌株，并进行体外实验、动物实验和人体实验，以证明其安全有效性。在这个环节中，企业需要投入大量的研究开发资源，以确保他们的产品具有优越的品质和较强的竞争力。

其次是益生菌菌株的发酵生产环节，这是上游环节的延伸。在这个环节中，菌种的分离，液体或固体发酵，菌体制备、浓缩、冷冻、干燥等多个步骤需要被精细地控制。因此，制定适宜的生产方案以确保生产出符合质量标准要求的产品是关键。

上游环节的主要目的是为下游提供高品质的益生菌

原料。下游包括2个环节，即配方产品的制作和品牌推广销售。在配方产品的制作环节，企业需要将稀释的原菌粉与其他原料成分混合制成成品。虽然技术难度不大，但剂型和配方的不同仍会影响益生菌活性和有效性。在品牌推广销售环节，企业需要根据品牌定位和目标人群，将产品投放到相应的渠道中。这个环节的成功与否往往取决于企业的市场营销策略和执行能力。

总的来说，上游环节是益生菌产业的核心，因为高品质的益生菌菌株是生产高品质益生菌产品的基础。上游环节的竞争十分激烈，包括益生菌菌株和生产技术的专利争夺。此外，上游环节对企业的要求是多方面的，因此具有较高的知识产权壁垒、技术壁垒、准入壁垒和资金壁垒。在这个竞争激烈的环境中，企业需要不断地投入研发资源，加强技术创新，以确保产品的市场竞争力。

（杨祺　令桢民*）

参考文献

[1]SCHNADOWER D,O'CONNELL K J,VANBUREN J M,et al. Association between diarrhea duration and severity and probiotic efficacy in children with acute gastroenteritis[J]. Official journal of the american college of gastroenterology,2021,116(7):1523-1532.

[2]SCHNADOWER D, SAPIEN RE, CASPER T C, et al. Association between age,weight,and dose and clinical response to probiotics in children with acute gastroenteritis[J]. The journal of nutrition,2021,151(1):65-72.

[3]GROELE L,SZAJEWSKA H,SZALECKI M,et al. Lack of effect of *Lactobacillus rhamnosus* GG and *Bifidobacterium Lactis* BB12 on beta-cell function in children with newly diagnosed type 1 diabetes:a randomised controlled trial[J]. BMJ open diabetes research and care,2021,9(1):1-10.

[4]VELAYATI A,KAREEM I,SEDAGHAT M,et al. Does symbiotic supplemen-

tation which contains bacillus coagulans *Lactobacillus Rhamnosus*, *Lactobacillus Acidophilus* and *Fructooligosaccharide* has favourite effects in patients with type-2 diabetes? a randomised, double-blind, placebo-controlled trial[J]. Archives of physiology and biochemistry, 2021, 129(6): 1–8.

[5]ZENG J, WANG C T, ZHANG F S, et al. Effect of probiotics on the incidence of ventilator-associated pneumonia in critically ill patients: a randomized controlled multicenter trial[J]. Intensive care med, 2016, 42(6): 1018–1028.

[6]REZAZADEH L, ALIPOUR B, JAFARABADI M A, et al. Daily consumption effects of probiotic yogurt containing *Lactobacillus acidophilus* LA5 and *Bifidobacterium lactis* BB12 on oxidative stress in metabolic syndrome patients[J]. Clinical nutrition ESPEN, 2021, 41: 136–142.

[7]CASTRO-HERRERA V M, FISK H L, WOOTTON M, et al. Combination of the probiotics *Lacticaseibacillus rhamnosus* GG and *Bifidobacterium animalis subsp. Lactis*, BB12 has limited effect on biomarkers of immunity and inflammation in older people resident in care homes: results from the probiotics to reduce infections in care home residents randomized, controlled trial[J]. Frontiers in immunology, 2021, 12: 1–10.

[8] TURCO R, RUSSO M, BRUZZESE D, et al. Efficacy of a partially hydrolysed formula, with reduced lactose content and with *Lactobacillus Reuteri* DSM 17938 in infant colic: a double blind, randomised clinical trial[J]. Clinical nutrition, 2021, 40(2): 412–419.

[9]MARTÍ M, SPRECKELS J E, RANASINGHE P D, et al. Effects of *Lactobacillus Reuteri* supplementation on the gut microbiota in extremely preterm infants in a randomized placebo-controlled trial[J]. Cell reports medicine, 2021, 2(3): 1–16.

[10]MAYA-BARRIOS A, LIRA-HERNANDEZ K, JIMÉNEZ-ESCOBAR I, et al. *Limosilactobacillus reuteri* ATCC PTA 5289 and DSM 17938 as adjuvants to improve evolution of pharyngitis/tonsillitis in children: randomised controlled trial[J]. Beneficial microbes, 2021, 12(2): 137–145.

[11]DE OLIVEIRA A M, LOURENCO T G, COLOMBO A P. Impact of systemic probiotics as adjuncts to subgingival instrumentation on the oral-gut microbiota associated with periodontitis: a randomized controlled clinical trial[J]. Journal of periodontology, 2022, 93(1): 31–44.

[12]CALANDRE E P, HIDALGO-TALLON J, MOLINA-BAREA R, et al. The probiotic VSL#3 does not seem to be efficacious for the treatment of gastrointestinal symptomatology of patients with fibromyalgia: a randomized, double-blind, placebo-controlled clinical trial[J]. Pharmaceuticals, 2021, 14(10): 1–15.

[13]MÄKELÄ S M, FORSSTEN S D, KAILAJÄRVI M, et al. Effects of *Bifidobacterium Animalis ssp. lactis 420* on gastrointestinal inflammation induced by a nonsteroidal anti-inflammatory drug: a randomized, placebo-controlled, double-blind clinical trial[J]. British journal of clinical pharmacology, 2021, 87(12): 4625–4635.

[14] PELLONPERÄ O, VAHLBERG T, MOKKALA K, et al. Weight gain and body composition during pregnancy: a randomised pilot trial with probiotics and/or fish oil[J]. British journal of nutrition, 2021, 126(4): 541–551.

[15]WU C C, WONG L C, HSU C J, et al. Randomized controlled trial of probiotic PS128 in children with tourette syndrome[J]. Nutrients, 2021, 13(11): 1–12.

[16]CHEN H M, KUO P H, HSU C Y, et al. Psychophysiological effects of *Lactobacillus Plantarum* PS128 in patients with major depressive disorder: a preliminary 8-week open trial[J]. Nutrients, 2021, 13(11): 1–15.

[17]WICKENS K L, BARTHOW C A, MURPHY R, et al. Early pregnancy probiotic supplementation with *Lactobacillus Rhamnosus* HN001 may reduce the prevalence of gestational diabetes mellitus: a randomised controlled trial[J]. British journal of nutrition, 2017, 117(6): 804–813.

[18]WAUTERS L, SLAETS H, DE PAEPE K, et al. Efficacy and safety of spore-forming probiotics in the treatment of functional dyspepsia: a pilot randomised, double-blind, placebo-controlled trial[J]. The lancet gastroenterology & hepatology, 2021, 6(10): 784–792.

[19]MCGUCKIN M A, ERI R, SIMMS L A, et al. Intestinal barrier dysfunction in inflammatory bowel diseases[J]. Inflammatory bowel diseases, 2009, 5(1): 100–113.

[20]MADSEN K, CORNISH A, SOPER P, et al. Probiotic bacteria enhance murine and human intestinal epithelial barrier function[J]. Gastroenterology, 2001, 121(3): 580–591.

[21]LEBEER S, CLAES I, TYTGAT H L P, et al. Functional analysis of *Lactobacillus Rhamnosus* GG pili in relation to adhesion and immunomodulatory interactions

with intestinal epithelial cells[J]. Applied and environmental microbiology, 2012, 78(1): 185–193.

[22]YAN F, POLK D B. Probiotic bacterium prevents cytokine-induced apoptosis in intestinal epithelial cells[J]. Journal of biological chemistry, 2002, 277(52): 50959–50965.

[23]VAN BAARLEN P, TROOST F, VAN DER MEER C, et al. Human mucosal in vivo transcriptome responses to three *Lactobacilli* indicate how probiotics may modulate human cellular pathways[J]. Proceedings of the national academy of sciences, 2011, 108(supplement_1): 4562–4569.

[24]WANG H, GONG J, WANG W, et al. Are there any different effects of *Bifidobacterium*, *Lactobacillus* and *Streptococcus* on intestinal sensation, barrier function and intestinal immunity in PI-IBS mouse model?[J]. PLoS one, 2014, 9(3): 1–8.

[25]BERMUDEZ-BRITO M, PLAZA-DÍAZ J, MUÑOZ-QUEZADA S, et al. Probiotic mechanisms of action[J]. Annals of nutrition and metabolism, 2012, 61(2): 160–174.

[26]ORTEGA-ANAYA J, MARCINIAK A, JIMÉNEZ-FLORES R. Milk fat globule membrane phospholipids modify adhesion of *Lactobacillus* to mucus-producing Caco-2/Goblet cells by altering the cell envelope[J]. Food research international, 2021(146): 1–9.

[27]BUCK B L, ALTERMANN E, SVINGERUD T, et al. Functional analysis of putative adhesion factors in *Lactobacillus Acidophilus* NCFM[J]. Applied and environmental microbiology, 2005, 71(12): 8344–8351.

[28]SUN J, LE G W, SHI Y H, et al. Factors involved in binding of *Lactobacillus Plantarum* LP6 to rat small intestinal mucus[J]. Letters in applied microbiology, 2007, 44(1): 79–85.

[29]COLLADO M C, GUEIMONDE M, SANZ Y, et al. Adhesion properties and competitive pathogen exclusion ability of *Bifidobacteria* with acquired acid resistance[J]. Journal of food protection, 2006, 69(7): 1675–1679.

[30]ISIDRO R A, LOPEZ A, CRUZ M L, et al. The probiotic VSL#3 modulates colonic macrophages, inflammation, and microflora in acute trinitrobenzene sulfonic acid colitis[J]. Journal of histochemistry & cytochemistry, 2017, 65(8): 445–461.

[31] PRINCE T, MCBAIN A J, O'NEILL C A. *Lactobacillus Reuteri* protects epidermal keratinocytes from staphylococcus aureus-induced cell death by competitive exclusion[J]. Applied and environmental microbiology, 2012, 78(15): 5119–5126.

[32] HOJO K, NAGAOKA S, MURATA S, et al. Reduction of vitamin K concentration by salivary *Bifidobacterium* strains and their possible nutritional competition with *Porphyromonas Gingivalis*[J]. Journal of applied microbiology, 2007, 103(5): 1969–1974.

[33] MARTIN F P J, WANG Y, SPRENGER N, et al. Probiotic modulation of symbiotic gut microbial-host metabolic interactions in a humanized microbiome mouse model[J]. Molecular systems biology, 2008, 4(1): 1–15.

[34] ELLI M, ZINK R, RYTZ A, et al. Iron requirement of *Lactobacillus spp.* in completely chemically defined growth media[J]. Journal of applied microbiology, 2000, 88(4): 695–703.

[35] WOO J, AHN J. Probiotic-mediated competition, exclusion and displacement in biofilm formation by food-borne pathogens[J]. Letters in applied microbiology, 2013, 56(4): 307–313.

[36] WHITTENBURY R. Hydrogen peroxide formation and catalase activity in the lactic acid bacteria[J]. Journal of general and applied microbiology, 1964, 35(1): 13–26.

[37] PENG X, ZHU L, GUO J, et al. Enhancing biocompatibility and neuronal anti-inflammatory activity of polymyxin B through conjugation with gellan gum[J]. International journal of biological macromolecules, 2020, 147: 734–740.

[38] WLADYKA B, PIEJKO M, BZOWSKA M, et al. A peptide factor secreted by *Staphylococcus Pseudintermedius* exhibits properties of both bacteriocins and virulence factors[J]. Scientific reports, 2015, 5(1): 1–15.

[39] GRANDE BURGOS M J, PULIDO R P, AGUAYO M, et al. The cyclic antibacterial peptide enterocin as-48: isolation, mode of action, and possible food applications[J]. International journal of molecular sciences, 2014, 15(12): 22706–22727.

[40] SINGH V P. Recent approaches in food bio-preservation-a review[J]. Open veterinary journal, 2018, 8(1): 104–111.

[41] STILES M E, HOLZAPFEL W H. Lactic acid bacteria of foods and their cur-

rent taxonomy[J]. International journal of food microbiology,1997,36(1):1–29.

[42]LIAO S F,NYACHOTI M. Using probiotics to improve swine gut health and nutrient utilization[J]. Animal nutrition,2017,3(4):331–343.

[43]BUJALANCE C,JIMÉNEZ-VALERA M,MORENO E,et al. Lack of correlation between in vitro antibiosis and in vivo protection against enteropathogenic bacteria by probiotic *Lactobacilli*[J]. Research in microbiology,2014,165(1):14–20.

[44]NIELSEN D S,CHO G S,HANAK A,et al. The effect of bacteriocin-producing *Lactobacillus Plantarum* strains on the intracellular pH of sessile and planktonic *Listeria Monocytogenes* single cells[J]. International journal of food microbiology,2010,141:53–59.

[45]HASSAN M,KJOS M,NES I F,et al. Natural antimicrobial peptides from bacteria:characteristics and potential applications to fight against antibiotic resistance[J]. Journal of applied microbiology,2012,113(4):723–736.

[46]HASHEMZADEH F,RAHIMI S,TORSHIZI M A K,et al. Usage of green fluorescent protein for tracing probiotic bacteria in alimentary tract and efficacy evaluation of different probiotic administration methods in broilers[J]. Journal of agricultural science and technology,2015,17:345–356.

[47]CORR S C,LI Y,RIEDEL C U,et al. Bacteriocin production as a mechanism for the antiinfective activity of *Lactobacillus Salivarius* UCC118[J]. Proceedings of the national academy of sciences,2007,104(18):7617–7621.

[48] TRICK W E,SOKALSKI S J,JOHNSON S,et al. Effectiveness of probiotic for primary prevention of clostridium difficile infection:a single-center before-and-after quality improvement intervention at a tertiary-care medical center[J]. Infection control & hospital epidemiology,2018,39(7):765–770.

[49]DA SILVA T F,CASAROTTI S N,DE OLIVEIRA G L V,et al. The impact of probiotics,prebiotics,and synbiotics on the biochemical,clinical,and immunological markers,as well as on the gut microbiota of obese hosts[J]. Critical reviews in food science and nutrition,2021,61(2):337–355.

[50]FIJAN S. Influence of the growth of *Pseudomonas Aeruginosa* in milk fermented by multispecies probiotics and kefir microbiota[J]. Journal of public health,2015,4(1):1–6.

[51]ZOETENDAL E G, RAES J, VAN DEN BOGERT B, et al. The human small intestinal microbiota is driven by rapid uptake and conversion of simple carbohydrates [J]. International society for microbial ecology journal, 2012, 6(7): 1415–1426.

[52]ZHAO Y, LIU J, HAO W, et al. Structure-specific effects of short-chain fatty acids on plasma cholesterol concentration in male syrian hamsters[J]. Journal of agricultural food and chemistry, 2017, 65(50): 10984–10992.

[53]HARA H, HAGA S, AOYAMA Y, et al. Short-chain fatty acids suppress cholesterol synthesis in rat liver and intestine[J]. The journal of nutrition, 1999, 129(5): 942–948.

[54]DU Y, LI X, SU C, et al. Butyrate protects against high-fat diet-induced atherosclerosis via up-regulating ABCA1 expression in apolipoprotein E-deficiency mice [J]. British journal of pharmacology, 2020, 177(8): 1754–1772.

[55]DE BOER J F, KUIPERS F, GROEN A K. Cholesterol transport revisited: a new turbo mechanism to drive cholesterol excretion[J]. Trends in endocrinology & metabolism, 2018, 29(2): 123–133.

[56]JOYCE S A, MACSHARRY J, CASEY P G, et al. Regulation of host weight gain and lipid metabolism by bacterial bile acid modification in the gut[J]. Proceedings of the national academy of sciences, 2014, 111(20): 7421–7426.

[57]DALY J W, KEELY S J, GAHAN C G M. Functional and phylogenetic diversity of BSH and PVA enzymes[J]. Microorganisms, 2021, 9(4): 1–17.

[58]KENNY D J, PLICHTA D R, SHUNGIN D, et al. Cholesterol metabolism by uncultured human gut bacteria influences host cholesterol level[J]. Cell host & microbe, 2020, 28(2): 245–257.

[59]ROY T L, HASE E M, HUL M V, et al. *Dysosmobacter Welbionis* is a newly isolated human commensal bacterium preventing diet-induced obesity and metabolic disorders in mice[J]. Gut, 2022, 71(3): 534–543.

[60]KOHDA Y, MATSUMURA H. Obesity-related hypertension and enhanced plasma orexin-A level are attenuated by the consumption of thiamine water in diabetic rats under cerebral oxidative stress conditions[J]. Fundamental toxicological sciences, 2019, 6(9): 383–390.

[61]HEMPEL S, NEWBERRY S J, MAHER A R, et al. Probiotics for the preven-

tion and treatment of antibiotic - associated diarrhea: a systematic review and meta - analysis[J]. Journal of the american medical association, 2012, 307(18): 1959–1969.

[62]MA Y, YANG JY, PENG X, et al. Which probiotic has the best effect on preventing Clostridium difficile-associated diarrhea? a systematic review and network meta-analysis[J]. Journal of digestive diseases, 2020, 21(2): 69–80.

[63]MEKONNEN S A, MERENSTEIN D, FRASER C M, et al. Molecular mechanisms of probiotic prevention of antibiotic - associated diarrhea[J]. Current opinion in biotechnology, 2020, 61: 226–234.

[64]JAFARNEJAD S, SHAB-BIDAR S, SPEAKMAN J R, et al. Probiotics reduce the risk of antibiotic - associated diarrhea in adults (18–64 years) but not the elderly (>65 years)[J]. Nutrition in clinical practice, 2016, 31(4): 502–513.

[65]SZAJEWSKA H, KOŁODZIEJ M. Systematic review with meta-analysis: *Lactobacillus Rhamnosus* GG in the prevention of antibiotic - associated diarrhoea in children and adults[J]. Alimentary pharmacology & therapeutics, 2015, 42(10): 1149–1157.

[66]SZAJEWSKA H, KOLODZIEJ M. Systematic review with meta-analysis: *Saccharomyces Boulardii* in the prevention of antibiotic-associated diarrhoea[J]. Alimentary pharmacology & therapeutics, 2015, 42(7): 793–801.

[67]PATRO-GOLAB B, SHAMIR R, SZAJEWSKA H. Yogurt for treating antibiotic - associated diarrhea: systematic review and meta - analysis[J]. Nutrition, 2015, 31(6): 796–800.

[68]DACQUAY L C, TSANG D, CHAN D, et al. *E.coli* Nissle increases transcription of flagella assembly and formate hydrogenlyase genes in response to colitis[J]. Gut microbes, 2021, 13(1): 1–14.

[69]LIU H Y, GIRAUD A, SEIGNEZ C, et al. Distinct B cell subsets in Peyer's patches convey probiotic effects by *Limosilactobacillus Reuteri*[J]. Microbiome, 2021, 9(1): 1–18.

[70]WANG L, LIAO Y, YANG R, et al. An engineered probiotic secreting Sj16 ameliorates colitis via ruminococcaceae/butyrate/retinoic acid axis[J]. Bioengineering & translational medicine, 2021, 6(3): 1–12.

[71]DORON I, MESKO M, LI X V, et al. Mycobiota-induced IgA antibodies regu-

late fungal commensalism in the gut and are dysregulated in Crohn's disease[J]. Nature microbiology,2021,6(12):1493-1504.

[72]RINGEL Y,CARROLL I M. Alterations in the intestinal microbiota and functional bowel symptoms[J]. Gastrointestinal endoscopy clinics of north america,2009,19(1):141-150.

[73]SALONEN A,DE VOS W M,PALVA A. Gastrointestinal microbiota in irritable bowel syndrome:present state and perspectives[J]. Microbiology,2010,156(11):3205-3215.

[74]MOAYYEDI P,FORD A C,TALLEY N J,et al. The efficacy of probiotics in the treatment of irritable bowel syndrome:a systematic review[J]. Gut,2010,59(3):325-332.

[75]O'MAHONY L,MCCARTHY J,KELLY P,et al. *Lactobacillus* and *Bifidobacterium* in irritable bowel syndrome:symptom responses and relationship to cytokine profiles[J]. Gastroenterology,2005,128(3):541-551.

[76]DE PALMA G,COLLINS S M,BERCIK P. The microbiota-gut-brain axis in functional gastrointestinal disorders[J]. Gut microbes,2014,5(3):419-429.

[77]NOBAEK S,JOHANSSON M L,MOLIN G,et al. Alteration of intestinal microflora is associated with reduction in abdominal bloating and pain in patients with irritable bowel syndrome[J]. The American journal of gastroenterology,2000,95(5):1231-1238.

[78]MCFARLAND L V,DUBLIN S. Meta-analysis of probiotics for the treatment of irritable bowel syndrome[J]. World J gastroenterol,2008,14(17):2650-2661.

[79]CARROLL I M,RINGEL-KULKA T,KEKU T O,et al. Molecular analysis of the luminal-and mucosal-associated intestinal microbiota in diarrhea-predominant irritable bowel syndrome[J]. American journal of physiology-gastrointestinal and liver physiology,2011,301(5):799-807.

[80]ZOCCO M A,DAL VERME L Z,Cremonini F,et al. Efficacy of *Lactobacillus* GG in maintaining remission of ulcerative colitis[J]. Alimentary pharmacology & therapeutics,2006,23(11):1567-1574.

[81]KRUIS W,FRIČ P,POKROTNIEKS J,et al. Maintaining remission of ulcerative colitis with the probiotic *Escherichia coli* Nissle 1917 is as effective as with stan-

dard mesalazine[J]. Gut,2004,53(11):1617-1623.

[82]SHEN J,ZUO Z X,MAO A P. Effect of probiotics on inducing remission and maintaining therapy in ulcerative colitis,Crohn's disease,and pouchitis:meta-analysis of randomized controlled trials[J]. Inflammatory bowel disease,2014,20(1):21-35.

[83]PERSBORN M,GERRITSEN J,WALLON C,et al. The effects of probiotics on barrier function and mucosal pouch microbiota during maintenance treatment for severe pouchitis in patients with ulcerative colitis[J]. Alimentary pharmacology & therapeutics,2013,38(7):772-783.

[84]HOLUBAR S D,CIMA R R,SANDBORN W J,et al. Treatment and prevention of pouchitis after ileal pouch-anal anastomosis for chronic ulcerative colitis[J]. Cochrane database of systematic reviews,2010,5(5):1-31.

[85]SALARIS C,SCARPA M,ELLI M,et al. *Lacticaseibacillus Paracasei* DG enhances the lactoferrin anti-SARS-CoV-2 response in Caco-2 cells[J]. Gut Microbes,2021,13(1):1-12.

[86]WU C,XU Q,CAO Z,et al. The volatile and heterogeneous gut microbiota shifts of COVID-19 patients over the course of a probiotics-assisted therapy[J]. Clinical and translational medicine,2021,11(12):1-5.

[87]XU J,REN Z,CAO K,et al. Boosting vaccine-elicited respiratory mucosal and systemic covid-19 immunity in mice with the oral *Lactobacillus Plantarum*[J]. Frontiers in nutrition,2021,8:1-11.

[88]CHEN G Y. The role of the gut microbiome in colorectal cancer[J]. Clinics in colon and rectal surgery,2018,31(3):192-198.

[89]GORI S,INNO A,BELLUOMINI L,et al. Gut microbiota and cancer:how gut microbiota modulates activity,efficacy and toxicity of antitumoral therapy[J]. Critical reviews in oncology/hematology,2019,143:139-147.

[90]SI W,LIANG H,BUGNO J,et al. *Lactobacillus Rhamnosus* GG induces cGAS/STING-dependent type I interferon and improves response to immune checkpoint blockade[J]. Gut,2022,71(3):521-533.

[91]CANALE FP,BASSO C,ANTONINI G,et al. Metabolic modulation of tumours with engineered bacteria for immunotherapy[J]. Nature,2021,598(7882):662-666.

[92]ALEXANDER J L, WILSON I D, TEARE J, et al. Gut microbiota modulation of chemotherapy efficacy and toxicity[J]. Nature reviews gastroenterol & hepatol, 2017, 14(6):356–365.

[93]PANEBIANCO C, PISATI F, ULASZEWSKA M, et al. Tuning gut microbiota through a probiotic blend in gemcitabine-treated pancreatic cancer xenografted mice [J]. Clinical and translational medicine, 2021, 11(11):1–7.

[94]ZHU H, CAO C, WU Z, et al. The probiotic *L. casei* Zhang slows the progression of acute and chronic kidney disease[J]. Cell metabolism, 2021, 33(10):1926–1942.

[95]PINTO-SANCHEZ M I, HALL G B, GHAJAR K, et al. Probiotic *Bifidobacterium Longum* NCC3001 reduces depression scores and alters brain activity: a pilot study in patients with irritable bowel syndrome[J]. Gastroenterology, 2017, 153(2): 448–459.

[96]MAROTTA A, SARNO E, DEL CASALE A, et al. Effects of probiotics on cognitive reactivity, mood, and sleep quality[J]. Frontiers in psychiatry, 2019, 10:1–11.

[97]HEMERT S, VERWER J, SCHÜTZ B. Clinical studies evaluating effects of probiotics on parameters of intestinal barrier function[J]. Advances in microbiology, 2013, 3(2):212–221.

[98]TSAI Y T, CHENG P C, PAN T M. The immunomodulatory effects of lactic acid bacteria for improving immune functions and benefits[J]. Applied microbiology biotechnology, 2012, 96(4):853–862.

[99]NG S C, HART A L, KAMM M A, et al. Mechanisms of action of probiotics: recent advances[J]. Inflammatory bowel diseases, 2009, 15(2):300–310.

[100]SHARMA R, KAPILA R, DASS G, et al. Improvement in Th1/Th2 immune homeostasis, antioxidative status and resistance to pathogenic *E. coli* on consumption of probiotic *Lactobacillus Rhamnosus* fermented milk in aging mice[J]. Official journal of the american aging association, 2014, 36(4):1–17.

第4章 肠道修复的概念及其应用

2021年9月29日下午，中共中央政治局就加强我国生物安全建设进行第三十三次集体学习。中共中央总书记习近平在主持学习时强调："要把优秀传统理念同现代生物技术结合起来，中西医结合、中西药并用，集成推广生物防治、绿色防控技术和模式，协同规范抗菌药物使用，促进人与自然和谐共生。"

4.1 食源性污染

4.1.1 食源性污染的现状及健康风险

食品安全作为人类生命健康的基础，对于提高人类生活质量起着决定性的作用，然而近几年随着食品行业的发展，食品安全危机频发，食品安全风险广泛影响到人民生活质量和国家社会稳定发展。据WHO统计，全球每年有6亿人（几乎每10人中就有1人）因食用不卫生的食品而患病，其中42万人因此死亡，而不安全的食物给低收入和中等收入国家造成的生产力和医疗费用损失高达1100亿美元。

食源性疾病是人体摄入污染食物或饮品后，各种具有感染性或毒性的致病因子引发的一系列疾病[1]，霍乱弧菌（*Vibrio cholerae*）、沙门氏菌、肠出血性大肠杆菌（*Enterohemorrhagic escherichia coli*）、诺如病毒等都是最常见的食源性病原体。但是自然产生的毒素和环境污染物同样是人体健康相关的研究热点。自然产生的毒素包括霉菌毒素、真菌毒素、海洋生物毒素、生氰苷和有毒蘑菇产生的毒素等。霉菌毒素在玉米或谷类等主食中含量较高，例如谷物上的霉菌会产生黄曲霉毒素和赭曲霉毒素，长期接触可能影响免疫系统和正常发育或导致癌症等。一些有毒的鱼类、贝类和毒蘑菇等动植物也是导致食源性疾病的食物之一。对于环境污染物而言，二噁英和多氯联苯作为持久性有机污染物，是工业生产和废物焚烧的有害副产品，常见于各种自然环境，会在动物食物链中积累，可能会损害免疫系统、干扰荷尔蒙、引起生殖和发育问题，甚至引发癌症。而且城市化建设过程中会产生大量废水、废气、废渣，它们均严重污染空气、水和土壤，例如重金属铅、镉和汞等均可造成神经系统及肾损害[2]。抗生素耐药菌或基因会破坏肠道菌群的稳定，改变菌群组成，也会对人体健康产生影响[3]。由于农产品的病虫害而违规滥用高毒农药化肥、激素等，使有毒有害物质超标残留而造成食物中毒。且违规滥用非法添加剂、不合格包装材料、非食用的化工类化合物等严重违法行为，在加工食品或保健品行业也屡见不鲜。表4-1统计了常见食源性病原体及其对医疗和经济的影响。

表4-1 常见食源性病原体及其影响[4]

食源性危害	常见的病原体	发病数/例	死亡数/例
细菌	沙门氏菌、弧菌、出血性大肠杆菌、志贺氏菌、李斯特菌、布鲁氏菌、弯曲杆菌等	359747420	272554
病毒	诺如病毒、甲型肝炎等	138513782	120814
原生动物类	内阿米巴、隐球菌、弓形虫等	77462734	6242
寄生虫	卵虫、线虫、吸虫等	26063664	90261
化学物质	黄曲霉毒素、氰化物、二噁英、重金属等	217632	19712

4.1.2 现有治疗方法及肠道修复的提出

人体肠道内栖息着1000种以上的细菌，总数接近1.0×10^{13}～1.0×10^{14}个，肠道内的大部分细菌在人体结肠内定植，其中每克肠内容物中细菌含量达到1.0×10^{12} CFU[5]。这些微生物组成的整体被称作肠道菌群。肠道微生态系统是机体内最庞大、最重要的微生态系统，对人体的健康与营养起着重要作用，是激活和维持肠道生理功能的关键因素。在正常的情况下，肠道内的微生物相互制约，保持相对平衡，种群维持在一定的比例范围内。当外源异物入侵或机体抵抗力低下时，都会对肠道菌群产生一定的破坏，使肠内优势菌群失调[6]，进而引起一系列生理反应。

益生菌和活微生物在补充足量时对人体健康有益[7]，被认为是防止食源性污染物有前途的工具。例如微生态调节剂双歧杆菌是人体内的优势菌群，作为肠道内的常驻菌，口服能增加体内益生菌的数量并抑制致病菌的生长[8]。双歧三联、四联杆菌可有效抑制出血性大肠杆菌、痢疾志贺菌等肠道致病菌的生长繁殖，在肠道内建立生物性保障屏障，使之不受胃酸、胆汁等的影响，摄入后能顺利进入肠道、迅速生长繁殖，促进双歧杆菌、乳酸杆菌等益生菌的生长，抑制有害菌及腐败菌的生长和繁殖[9]。另外，寻找改善疾病的天然细菌的另一种方法是使用合成生物学工具，即让一种经过基因改造的细菌去执行特定功能以获得诊断或治疗益处[10]。因此，开发工程菌株，可以像典型的益生菌一样，扮演肠道定植者的角色，作为肠道微生态调节剂，保护肠道免受环境污染物的侵害，故开发工程菌株是一种潜在的环保、有效的解决方案。将肠道中污染物的修复过程称为“肠道修复”。

4.2 肠道修复在缓解食源性污染方面的应用现状

4.2.1 重金属与疾病

重金属污染正在成为一个日益严重的全球性环境问题[11]，人类活动以及迅速的工业化和城市化都将导致大面积的土壤和水资源受到重金属污染。据统计，Pb、Cr和Hg包含于全球十大污染物之中，因其极难降解，低剂量即可经食物链生物富集对人类健康构成了巨大威胁[12]。重金属中毒依据其暴露的时长及暴露剂量分为急性中毒和慢性中毒。高剂量重金属暴露诱发的急性中毒常常反应迅速、症状明显，而短期低于中毒阈值的重金属暴露症状不明显，不会导致急性中毒反应，但是重金属在靶器官的持续累积易造成严重的慢性损伤，逐渐表现出重金属中毒的症状并慢性迁延加重，如引起炎症[13]，诱发氧化应激[14，15]、各种组织病变[16]和肠道疾病[17]，并增加患癌和其他疾病的风险。

近年来，人们提出了许多技术以应对土壤重金属污染，如植物修复、物理化学修复和生物学方法等。与传统的物理化学修复相比，植物修复由于经济环保，已被普遍用于重金属污染土壤的净化[18]。然而，大部分超富集植物往往只能在单一重金属污染的环境中生长，故植物修复对于复合污染治理无法有效应对。此外，大部分超富集植物的生长周期较长，生物量相对较小，这限制了它们在短时间内大量吸收和固定重金属的能力[19]。在不久的将来，人类将不可避免地接触到受重金属污染的食物，特别是在发展中国家。因此，迫切需要寻求实用的方法来保护人类免受重金属的侵害。

面对上述重金属污染治理需求，微生物修复技术应运而生，已成为重金属生物修复的一种环境友好的替代方法。例如乳酸杆菌等耐重金属的细菌物种可用于重金属生物修复[20，21]。已有研究发现口服植物乳杆菌可以促进小鼠从粪便排出重金属并通过肠道屏障抑制吸收保护其免受Cd、Pb、Cu和Al的毒性[22-25]。

肠道微生物群作为人类第二大基因组，在维护肠上皮细胞的稳态和调节代谢等方面发挥着十分重要的作用[26]。研究发现，肠道微生物组在抵御重金属暴露方面起着至关重要的作用[27]。大多数重金属在被摄入后进入肠道，可以被某些微生物吸收，阻止它们被肠道和其他组织吸收[28]。例如，凝结芽孢杆菌R11可以维持Pb暴露小鼠肠道绒毛的健康，并减少Pb暴露引起的肠道损伤和伤害[29]。然而，天然肠道微生物通常没有重金属特异性结合蛋白，因此开发具有金属特异性的重组菌株，将有利于解决金属暴露引起的生物健康问题。

4.2.2 肠道菌群缓解重金属对机体损害的研究进展

（1）益生菌酸奶通过改变肠道微生物群的组成和代谢降低重金属Cu和Ni的水平

益生菌被定义为活的微生物，当以足够的量施用时，会对胃肠道和免疫系统产生有益作用[30]，还可以预防糖尿病[31]、骨质疏松症、抑郁和焦虑，以及降低血压[32-35]，能有效减少有害物质在宿主中的积累，如有毒金属、抗生素和杀虫剂等的积累[36]。首次使用益生菌鼠李糖乳杆菌作为预防重金属积累干预剂的人体试验发现，益生菌在预防成人血液中有毒金属水平升高方面具有功效[37]。同样，另一项人体试验表明，食用长达8周的植物乳杆菌会显著降低矿区附近城镇居民血液中Cd的水平[38]。

研究显示，活性氧的过量产生损害了肠道微生物的正常功能和肠道屏障[39, 40]。一项关于As暴露小鼠的肠道微生物群的宏基因组研究表明，参与氧化应激反应的基因丰度显著上调，表明活性氧升高和肠道微生物功能受损[41]。益生菌的抗氧化能力对于维持氧化还原稳态至关重要。例如，小球菌属和乳酸杆菌通过谷胱甘肽过氧化物酶和过氧化氢酶减弱野生动物和人类中农药诱导活性氧的产生[42, 43]，乳酸杆菌通过其抗氧化活性抑制D-半乳糖诱导的小鼠氧化应激和肠道微生物群失调[44-46]。此外，从牦牛乳中分离出的两种益生菌菌株，显著减轻了小鼠的铬酸盐毒性[19]。所以，将益生菌作为一种修复人类重金属毒性的治疗药物，可以通过益生菌介导的抗氧化能力来维持肠道氧化还原稳态，保护肠道结构稳定和代谢正常，以增强肠道微生物对重金属的抵抗力[11]。

基于此，Feng等[47]研究发现饮用益生菌酸奶可显著降低血液中Cu和Ni的水平。变形菌门作为肠道生态失调的指标，酸奶干预后其丰度显著降低，使机体局部和全身的炎症得以缓解[48-50]。放线菌群中的革兰氏阳性菌（如芽孢杆菌、肠球菌、乳酸菌、双歧杆菌等），由于其细胞壁中肽聚糖和磷壁酸含量高，对金属有较强的吸附和去除能力[21, 51, 52]，使用益生菌酸奶后，该属的宏基因组物种显著富集，增加了肠道微生物群排泄Cu和Ni的潜力。此外，肠道屏障功能的增强和肠道通透性的降低，使得异种杆菌、乳酸杆菌[53]和双歧杆菌等益生菌的丰度增加。通过小鼠实验比较GR-1和Vc的修复效果发现，益生菌的抗氧化能力是改善重金属诱导毒性的主要机制[23, 54]。益生菌GR-1在肠道中定植，对于改善小鼠Cu毒性和维持肠道微生物群稳态方面表现出长期作用，通过益生菌修复了小鼠肠道微生物菌群的基本结构。所以肠道微生物群的完整性在限制重金属摄取方面具有重要作用，GR-1通过减轻氧化应激和保护肠道微生物功能减轻了Cu和Ni对小鼠的毒性。

综上，使用具有强大抗氧化能力的益生菌对抗重金属毒性似乎是合理的。考虑到处于重金属暴露风险中的人群广泛，基于益生菌的肠道修复是重金属解毒可行且有前途的干预策略。为了增强益生菌的修复效果，与益生元（重要的健康促进剂[55]）共同给药可以考虑用于未来的应用。

（2）工程菌可减轻Pb^{2+}诱导的氧化应激和肠道组织改变

Pb^{2+}是一种对人类健康有害的环境风险物质，可导致不可逆转的神经损伤，也会造成肾、心血管和消化系统等的损害[56, 57]。环境中Pb^{2+}的存在主要与人造产品有关，如含铅汽油和油漆、铅酸电池、柴油机尾气[58]，Pb^{2+}最终通过自然循环排放到水生生态系统中[59]，若处理不当，会在海洋生物中富集，被人体食用后将会对人类健康造成潜在威胁。据报道，长江中下游泥鳅中Pb^{2+}质量分数已达到10.1 mg/kg，比法定限值高出数倍[60]。如今，Pb^{2+}污染正在世界范围内蔓延，因此有必要找到一种安全有效的方法来减轻鱼类的铅污染。

ACBP1、a3DIV和PbBD蛋白已被证明具有Pb^{2+}吸附能力[61, 62]，其中PbBD蛋白吸附效率较好，因此，利用冰核蛋白（ice nucleation protein，INP）[63]构建在大肠杆菌表面显示PbBD的工程菌株，通过蛋白沉淀技术将Pb^{2+}特异性表面结合蛋白耦合在工程菌表面，研究发现其最大吸附效率可以达到73%（图4-1）[64]。肠道是环境来源铅的主要吸收部位[65]，将大肠杆菌PbBD引入草鱼肠道，草鱼体内Pb^{2+}的浓度显著降低，粪便中Pb^{2+}的残留量增加了76%，大肠杆菌PbBD有效减少了组织和肠道中铅的积累并且促进了铅的排泄过程。同时发现大肠杆菌PbBD能够在草鱼的肠道中成功存活和定植，减轻了Pb^{2+}暴露引起的肠道氧化应激和组织学变化以及肠道菌群的失调。肠道既是消化器官，也是免疫系统的一部分[66]，一般来说，重金属会损害肠绒毛并破坏黏液物质，从而影响肠道健康。不过，大肠杆菌PbBD的引入降低了Pb^{2+}在草鱼肠道中的暴露水平，从而降低了Pb^{2+}对肠细胞的毒性作用，使肠道中的黏液物质继续执行其与免疫、润滑、消化和肠道pH调节相关的功能。另外，Pb^{2+}暴露会导致一些潜在病原体（如志贺氏菌和拟杆菌）的过度生长，抑制肠道益生菌（如鲸杆菌属和芽殖杆菌属）的生长，并且会改变微生物的整体群落结构，特别是志贺氏菌，它是引起肠道感染的主要病原体之一，肠道中致病菌的过度生长可能会影响核心微生物的生长，导致机体健康状况不佳[67]。将工程大肠杆菌PbBD引入肠道后，可以通过其去除肠道Pb^{2+}，限制肠道对Pb^{2+}的吸收，恢复了Pb^{2+}暴露引起肠道微生物群落的组成和多样性的变化[68]，从而维持健康的肠道微环境并增强肠道免疫反应[69]。

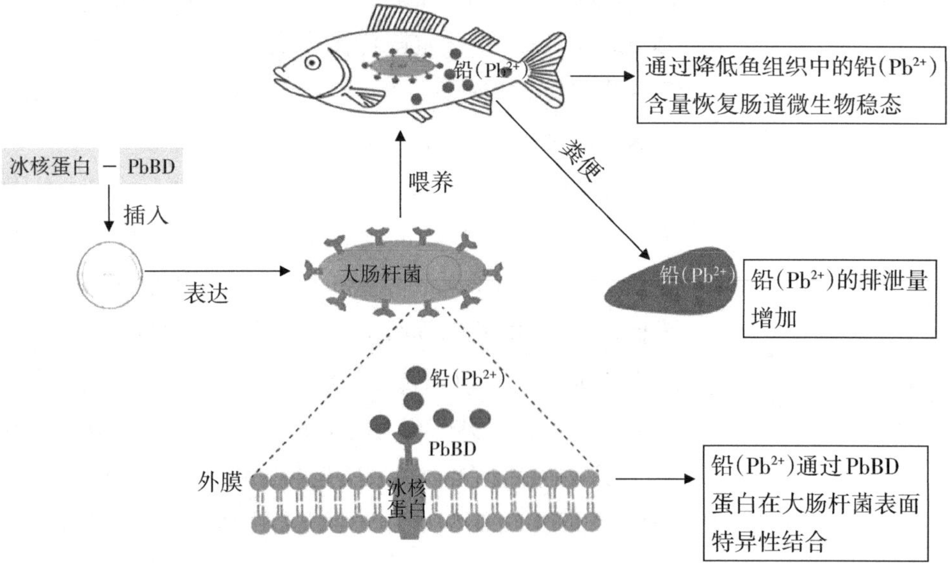

图4-1 大肠杆菌PbBD吸附Pb^{2+}示意图[64]

工程细菌可有效减轻Pb^{2+}对肠道的损害和毒性，因此，能够有效降低草鱼体内Pb^{2+}残留的工程菌可能是水生生物中铅去除的有效工具，将其作为饲料添加剂来使用有望解决Pb^{2+}污染引起的草鱼健康问题。这将是一种防止环境污染物进入高等生物体的新方法。

（3）工程菌株通过去除甲基汞在肠道发挥保护作用

汞是一种对人类、生态系统和环境具有剧毒的元素[70]。由于其在环境中可以长期存在，已被认为是全球主要的污染物之一[71]。汞可转化为具有神经毒性的甲基汞，甲基汞是食物链中汞的主要存在形式[72]。甲基汞的亲脂性使其可以在食物链中积累，进而引起严重的神经损伤，并增加人类和野生动物心脏病的发病率[73]。海鲜消费被认为是人类接触神经毒素甲基汞的主要途径[74]。鱼类体内的汞积累量是周围水生环境的1000倍[75]。据报道，地中海鱼类中汞的平均质量分数为0.54～2.22 μg/g[76]。一项基于水产品的研究发现，水产品中甲基汞的质量分数为4.3～303.6 ng/g[73]。因此，迫切需要找出一种减少鱼类中甲基汞积累的新方法。现如今，修复甲基汞的方法有很多，包括纳米技术、光降解和微生物降解等，但这些方法对鱼体内甲基汞的减少效果不佳，会导致多种健康问题，例如水俣病、发育不良和不良心血管[73-77]。相比之下，生物吸附是修复汞的有效方法。

Liu等[78]借助基因工程技术在大肠杆菌菌株W-1细胞表面显示一种新型甲基汞结合肽，肽序列以先前的报告为基础[79]，将其修改为CysLysCysLysCysLysCysLysCys（CL），冰核蛋白基因的N末端区域作为锚点[80]，使细胞显示CL肽，以此构建吸附甲基汞的表面工程大肠杆菌W1M/CL（图4-2）。研究发现表面工程大肠杆菌W1M/CL对甲基汞的生物吸附大于在细胞质中表达CL的工程大肠杆菌，其细胞表面显示增强了对甲基汞的吸附能力。该工程菌株对甲基汞表现出较高的亲和力和选择性，透射电子显微镜证实了甲基汞在其细胞膜上的积累，该工程菌株能有效去除96%以上12 μmol的甲基汞。用表面工程菌株W1M/CL饲喂鲫鱼后，鲫鱼肌肉中甲基汞浓度降低了约36.3%±0.7%，在肠道中与甲基汞结合的大肠杆菌W1M/CL可以通过粪便排泄，致使粪便中甲基汞浓度相对增加。肠道是鱼类摄入受汞污染的浮游植物后第一个被影响的器官[81]。因此，肠道微生物群（如益生菌）在维持鱼类的健康方面起着重要作用[82]。研究显示，甲基汞能在鲫鱼肠道中定植，其摄入会引起鲫鱼肠道细菌群落结构发生变化，使得部分机会致病菌（如施万菌属等）显著增加，并在肠道过度生长，这可能会影响鱼肠道核心微生物组成，进而引发疾病[83]。但是在肠道中引入的工程菌W1M/CL可以捕获甲基汞，阻止甲基汞被肌肉吸收，以此在肠道中发挥保护作用，使鱼类免受甲基汞毒性的影响。在另一项研究中，构建的工程菌株*E. coli* PC在保护了鱼肠道微生物多样性的同时，增加了汞离子的排泄、减少了汞离子在肌肉组织中的积累[84]；还发现这种表面工程细菌对汞离子表现出高容量和良好的选择性吸附，其他金属离子不会干扰汞的吸附效率[78]。

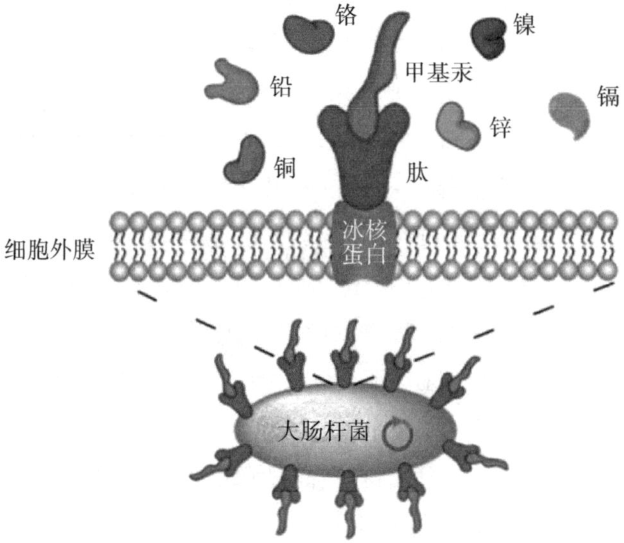

图4-2 表面工程大肠杆菌W1M/CL的构建示意图[78]

因此，使用工程菌解毒是减少鱼体内汞积累的理想策略。汞离子在工程菌细胞表面的吸附效率非常高，工程菌对汞离子的吸附具有成本效益，为去除生物体内汞的残留提供了一种新颖的设计策略。

(4) 部分益生菌通过响应Cr(Ⅵ)引起氧化应激防止Cr毒性

Cr是一种环境中常见的有毒重金属污染物，主要以三价(Ⅲ)或六价(Ⅵ)形式存在[85]。Cr(Ⅵ)很容易被活细胞吸收，是一种高度危险的Cr存在形式，会导致癌症和其他健康问题[86]。越来越多的证据表明，乳酸杆菌和潜在的益生菌可能有助于人类和小鼠模型中摄入的重金属的解毒。有研究发现植物乳杆菌菌株具有较强的Cr(Ⅵ)还原能力，有效减少了小鼠组织中Cr(Ⅵ)的积累，增强了粪便微生物的Cr(Ⅵ)还原能力[19]，有效减轻了摄入的镉、铅和铬酸盐对小鼠的毒性。所以Feng等[87]提出共生肠道微生物固有的Cr(Ⅵ)解毒基因的活性在Cr(Ⅵ)胁迫下受到抑制，但其可以被某些益生菌恢复，从而增强肠道微生物对Cr(Ⅵ)的抵抗能力。

从青藏高原酸奶中分离得到的酸乳片球菌BT36具有极强的抗氧化活性。研究发现小鼠暴露于Cr(Ⅵ)会诱导肝中的氧化应激，将酸乳片球菌BT36在小鼠肠道中定植后，可显著增加Cr(Ⅵ)的排泄，减少Cr(Ⅵ)在肝中的积累，从而降低肝氧化应激，减轻肝组织学损伤[19]。与此同时，BT36逆转了Cr(Ⅵ)诱导的肠道微生物组成结构和代谢活性的变化。具体而言，该菌恢复了788个基因的表达，其中包括34个固有的Cr(Ⅵ)修复相关基因，而且发现肠道微生物群中还存在新的Cr(Ⅵ)还原基因。因此，BT36可以调节肠道微生物群以响应Cr(Ⅵ)诱导的氧化应激并防止Cr(Ⅵ)发挥毒性作用。另一篇报道发现植物乳杆菌TW1-1也可以减少组织中Cr(Ⅵ)的积累[19]，减轻Cr(Ⅵ)暴露引起的氧化应激改变和组织病理学变化(图4-3)。在Cr(Ⅵ)诱导下给予TW1-1后，粪便中细菌的铬酸盐还原能力提高了1倍。TW1-1可以恢复部分因Cr(Ⅵ)暴露而引起肠道微生物群结构的改变。所以TW1-1作为肠道微生物群的调节剂，有助于铬酸盐减少并提供对Cr(Ⅵ)的保护。

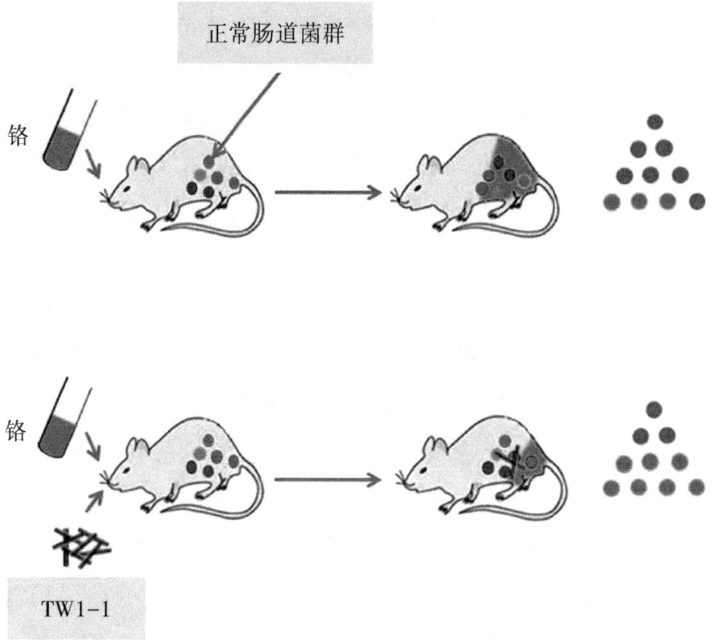

图4-3 TW1-1的摄入促进了铬的排泄[19]

4.3 肠道菌群与有机物降解

4.3.1 有机物与疾病

有机物是指含有碳化合物或者碳氢化合物及其衍化物的总称，所有的生命体都含有机化合物，如脂肪、酶、激素等。当生物体内的某个有机物含量超出或低于正常值时会引起疾病，如高尿酸血症、痛风等。环境中的持久性有机污染物（如甲基汞）、亲脂性卤代有机物（如二噁英、多氯联苯）在环境和人体中都具有抗降解性，并倾向于在食物链中进行生物积累，从而引起免疫系统障碍、认知及神经行为功能下降以及甲状腺功能受损等疾病，并且会增加患高血压、心血管疾病和糖尿病等慢性疾病的风险[88]。

4.3.2 通过肠道菌群降解有机物的研究进展

通过调节肠道菌群来降解生物体内有机物逐渐成为研究热点，这里主要介绍近年利用益生菌调节肠道菌群，从而改善高尿酸血症、睾丸损伤和预防酒精性肝病，以及利用工程菌降低动物粪便中抗生素含量的相关研究。

（1）发酵乳杆菌JL-3与GR-3可通过降解尿酸改善高尿酸血症

高尿酸血症是与痛风相关的主要代谢性疾病之一，其除了可发展为痛风，还可导致肾损伤，严重者会出现关节虫蚀样改变、肾功能损害，常伴发高脂血症、高血压病、糖尿病、动脉硬化及冠心病等[89]。2015年的荟萃分析显示，中国成人高尿酸血症的患病率高达13.3%（男性19.4%，女性7.9%），约有1.7亿高尿酸血症患者，而且大部分患者需要长期甚至终身服用降尿酸药物进行治疗，长期甚至终身服用降尿酸药物存在成本较高且具有一定的副作用等问题[90, 91]。因此，积极寻求安全、价廉、有效防治尿酸升高的策略具有重要临床和公共卫生意义。

浆水是中国西北地区的一种传统自然发酵食品，在其发酵过程中会产生大量氨基酸、B族维生素等物质，这些物质有助于机体的消化和吸收[92]。目前已有部分学者开展了对于浆水中微生物菌株分离以及鉴定工作，研究表明浆水中的优势菌为乳酸菌与酵母菌[93]，从浆水中分离出的棉子糖乳球菌Q7可以降解胆固醇且具有应用价值[94]。

近期，Wu等从浆水中分离出具有尿酸还原能力的候选益生菌菌株：发酵乳杆菌JL-3（图4–4），发现该菌株可以通过降解尿酸改善高尿酸血症[95]。研究人员前期通过对兰州市居民开展问卷调查，发现经常吃浆水面的人患痛风的可能性较小。研究人员从市售发酵的浆水中分离出20株具有尿酸抗性的菌株，在以尿酸为唯一碳氮源的培养基中培养，最终选择生长速度最快的JL-3菌株。后期以小鼠为动物模型，探讨JL-3对尿酸的降解作用时发现，JL-3能够有效降低高尿酸血症小鼠体内的尿酸水平，推测JL-3是通过降解肠道内尿酸，从而减少肠道内积累的尿酸含量的，这一过程也因此降低了通过肠道上皮重新进入循环的尿酸含量。JL-3还可以降低小鼠体内炎症标志物的含量以及氧化应激指标水平，如丙二醛、肌酐和血尿素氮。另外，发现体内尿酸水平高会改变小鼠肠道微生物群组成，而JL-3可调节高尿酸血症引起的肠道菌群失调，减小尿酸对微生物群变化的影响。因此，JL-3被认为可以调节肠道微生物群的结构与功能。

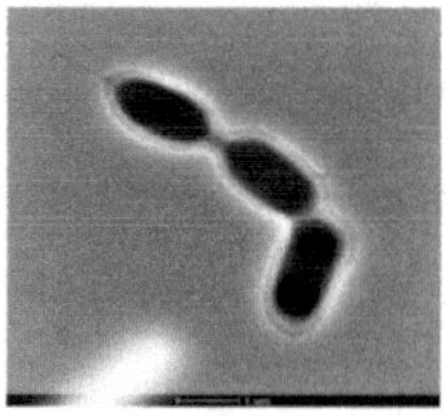

图4-4　JL-3的扫描电子显微图[95]

Zhao等研究发现，发酵乳杆菌GR-3也可以通过降解尿酸改善高尿酸血症[96]。发酵乳杆菌GR-3也是一种从自然发酵浆水中分离出的候选益生菌菌株，具有较强的尿酸降解功能，并且酸奶作为乳制品比胶囊更适合分布活性益生菌，因此该研究选择益生菌酸奶进行实验。共对120例血清尿酸水平高的患者进行了随机、双盲对照研究，志愿者们食用含有GR-3的益生菌酸奶或传统酸奶2个月后，进行样本收集及相关指标的测定（图4-5）。发现GR-3可以成功在肠道中定植并有效降解尿酸，从而降低血液中尿酸水平。通过食用GR-3益生菌酸奶，志愿者体内的血清尿酸水平与食用常规酸奶组的相比显著降低。GR-3还会显著调节肠道微生物区系的分类组成，进一步减少炎症，改善血清生化指标。通过降低血清中尿酸水平来改善胃肠道屏障功能，并且可以有效改善肾损伤。

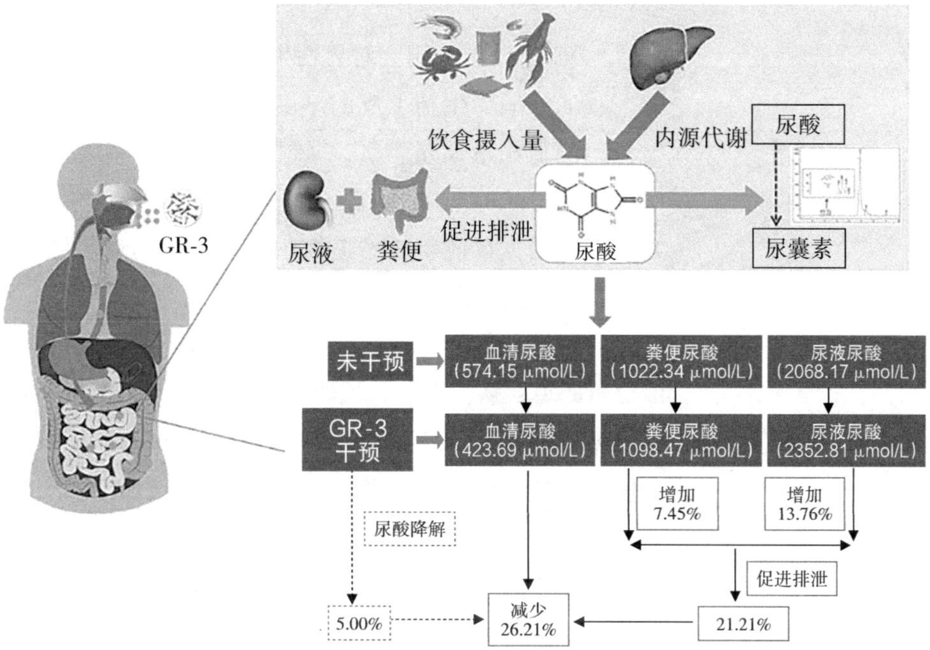

图4-5 GR-3菌株对人体中尿酸的降解途径[96]

目前，大多数轻度或中度高尿酸血症患者相比药物治疗，更容易接受食用酸奶的方式治疗，因此可以考虑将益生菌酸奶作为一种有效、经济且安全的辅助剂用于高尿酸血症的治疗。值得注意的是，这项研究并没有提供益生菌和全剂量药物的比较分析，还需要进一步的研究来确定益生菌是否比某些药物更具成本效益。与目前的治疗技术相比，益生菌可直接有效地缓解高尿酸血症，从而给高尿酸血症提供了一个有前景的治疗策略。

（2）植物乳杆菌TW1-1可减轻邻苯二甲酸二酯诱导的小鼠睾丸损伤

邻苯二甲酸二乙基己酯［di-(2-ethylhexyl)phthalate，DEHP］是生产柔性聚氯乙烯中最常用的邻苯二甲酸酯。日常生活中使用的食品包装袋、医疗器械、食品添加剂等产品中都含有DEHP。当DEHP进入环境循环后，会成为环境污染的主要来源，并且能够沿土壤—植物—食物链最终进入人体，对于人体具有潜在的致癌和诱变作用[97]。DEHP已被证明会对男性生殖道的发育产生不利影响，其通过减少生精上皮的厚度和细胞层数来诱导睾丸损伤，降低精液质量[98, 99]。据报道，邻苯二甲酸二乙酯（diethyl phthalate，DEP）是一种与DEHP类似的化合物，其可以改变大鼠的整体肠道细菌组成[100]，而近年已有研究证实肠道微生物群会参与调节睾丸内分泌功能[101]。

乳酸杆菌是一种益生菌，已被用于调节免疫系统和治疗胃肠道疾病。例如，植物乳杆菌CCFM8610能够保护小鼠免受镉毒性伤害，植物乳杆菌CCFM8661则可以减轻小鼠铅诱导的毒性[102]，并且补充罗伊氏乳杆菌会增加衰老小鼠的睾酮水平[103]。Tian等提出益生菌可用作修复策略，以减少环境污染物暴露引起的小鼠生理损伤[104]。从发酵乳制品中分离出的植物乳杆菌TW1-1具有抗炎和抗氧化应激活性，所以研究人员通过检查炎症标志物的含量、氧化应激指标的水平和肠道微生物群的组成，评估了植物乳杆菌TW1-1对DEHP诱导的成年雄性小鼠睾丸损伤的影响（图4-6）。

补充植物乳杆菌TW1-1可改善DEHP诱导的小鼠睾丸重量的下降情况，并且使过氧化氢酶、超氧化物歧化酶、脂质过氧化和丙二醛水平恢复到对照组小鼠水平，从而减轻小鼠体内的生殖毒性及氧化应激。植物乳杆菌TW1-1还能显著降低DEHP诱导的高浓度炎症标志物的含量，也抑制这些炎症标志物的基因表达，缓解DEHP暴露引起的全身和组织炎症以及肠道通透性增加现象。在门水平上，植物乳杆菌TW1-1可以逆转DEHP诱导的拟杆菌增加和厚壁菌减少的情况，从而恢复小鼠体内失衡的肠道微生物群。

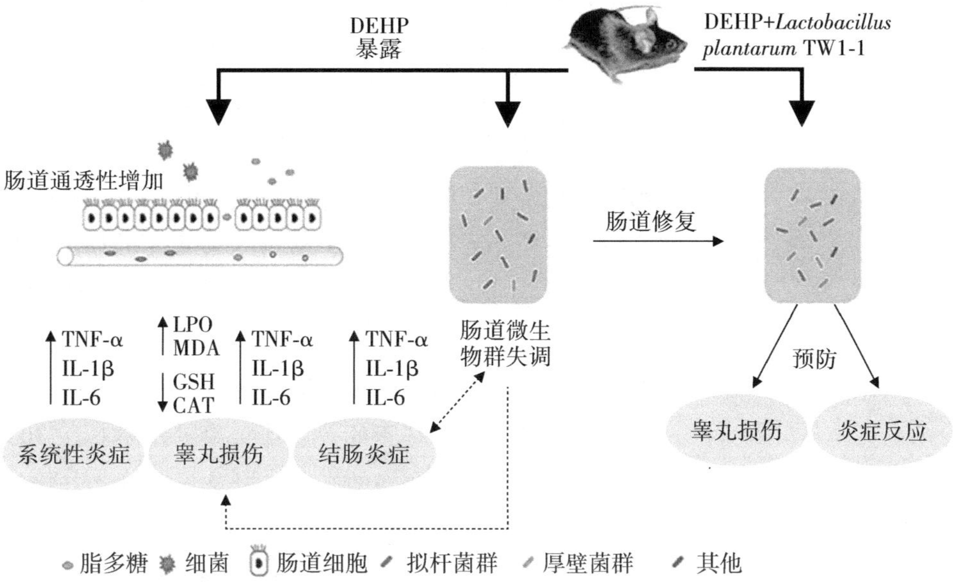

图4-6 植物乳杆菌TW1-1对DEHP诱导的睾丸损伤的保护途径[104]

该研究首次证实植物乳杆菌TW1-1可以通过调节肠道微生物群和减少炎症来预防DEHP诱导的小鼠睾丸损伤[104]。通过肠道修复调节生殖疾病相关的微生物群落为预防或治疗该类疾病提供了一种新的方法。

（3）短双歧杆菌ATCC15700通过调节小鼠肠道菌群来预防酒精性肝病

酒精性肝病（alcoholic liver disease，ALD）是由于长期大量饮酒导致的肝病，其临床症状各不相同，从无症状的肝脂肪变性到厌食、体重减轻、腹部不适、肝肿大等[105]。饮酒后超过90%的酒精会在肠道中吸收，其中的大部分在肝中代谢[106]。酒精及其代谢物乙醛能够直接损害肝蛋白，导致肝细胞损伤[107，108]。

据报道，饮酒会改变肠道微生物群的组成[109]，这一改变被称为肠道微生态失衡，在酒精性肝炎的发展中起着关键作用[110]。肠道微生态失衡会增加肠道通透性，导致肠道中的细菌和脂多糖容易向肝转移[111，112]。脂多糖能够通过Toll样受体4激活NF-κB的途径，导致肝炎症[113]。

一些双歧杆菌菌种可以预防酒精性肝病，如植物乳杆菌LC27和长双歧杆菌LC67可通过恢复受干扰的肠道微生物群来减轻酒精诱导小鼠的肝损伤[114]。但很少有研究确定这些菌株在调节免疫力和增强肠道屏障功能方面的能力。从婴儿粪便中分离出的短双歧杆菌ATCC15700被鉴定为具有抗炎活性的天然益生菌[115]。Tian等考虑到短双歧杆菌的多种活性，假设ATCC15700可以通过改变肠道微生物群调节免疫炎症反应和肠道屏障功能，从而缓解酒精性肝病[116]。基于此，设计实验研究ATCC15700对小鼠慢性酒精性肝损伤的影响，以及肠道微生物群在介导ATCC15700对酒精性肝病影响中的作用。

ATCC15700预处理会降低酒精诱导的肝脂肪变性、肝细胞增殖以及血清和肝炎症细胞因子数量，保护暴露于酒精的小鼠免受肝损伤。另外，通过调控TLR4-NF-κB途径可有效预防肝炎症，使TLR4和NF-κB p65在肝中的浓度恢复到正常水平。ATCC15700通过增强紧密连接蛋白的表达改善肠道屏障功能，以增加排便率，促进肠道蠕动。因此，ATCC15700对酒精性肝病的保护作用可能源于它对丁酸球菌、瘤胃球菌等肠道微生物群的调节，这些特定细菌已被证实参与ATCC15700对酒精暴露小鼠ALD的保护。

该研究结果将为今后临床研究和益生菌开发与应用提供理论依据。可以考虑将肠道微生物群中特定细菌或关键分类群的改变用于疾病预测，并且通过靶向益生菌调节肠道微生物群可能是酒精性肝病的一种有应用前景的治疗策略。

（4）生物工程菌株可降低动物粪便中的抗生素水平

抗生素是新兴有机污染物，会对人类健康和生态系统产生不利影响，已被世界卫生组织视为三大危险物之一[117，118]。抗生素被广泛用于牲畜的医疗和养殖中，但大量的抗生素在动物体内无法被代谢或吸收[119]，动物体内50%～90%的抗生素以原始或化合物形式排泄[120，121]。因此，动物粪便就成了抗生素污染的主要来源之一[117]。

目前，非生物和生物处理已被用于去除动物粪便中残留的抗生素[121，122]。焚烧作为一种有效技术，可被用于处理动物粪便[123]，然而这项技术由于会产生大量烟雾，在处理过程中可能造成空气污染[124]。除了焚烧技术之外，还有堆肥[125]、厌氧消化[126]等技术可以利用，但这些常规技术不足以完全去除粪便中的抗生素，例如残留的抗生素会减少堆肥期间嗜热期的持续时间并影响堆肥的效果[127，128]。因此，开发一种新的策略对于去除抗生素以及减少从粪便释放到环境中的抗生素至关重要[129]。据报道，动物肠道中的功能性细菌定植能够降解环境污染物，可以利用黑水虻幼虫有效及快速降解四环素的能力进行粪便处理，这种修复策略成本低，并且能够安全方便地去除污染物[130，131]。此外，还可以通过构建表面工程细菌，将一些功能性酶显示在细菌细胞表面，降解有机污染物，从而克服底物运输和酶纯化等限制[132，133]。

源自大肠杆菌质粒或铜绿假单胞菌的红霉素酯酶能够降解红霉素，而红霉素是使用最广泛的大环内酯类抗生素之一，并且由于其在环境中持久性已成为顽固污染物[134，135]。Liu等假设大肠杆菌细胞表面显示的红霉素酯酶可作为一种新的工程细菌在大肠定植以降低粪便中的红霉素水平[136]，在该研究中，红霉素酯酶通过冰成核蛋白显示在大肠杆菌细胞表面（图4-7），并将工程细菌饲喂给小鼠，以评估动物粪便中红霉素的去除情况。

我们利用重叠PCR融合*InaK-N*和*ereA*基因构建红霉素酯酶细胞表面展示质粒。用HindIII和XhoI酶切InaK-N-ereA融合片段，插入到质粒pET23b中，得到重组质粒pET23ereA，并将其转化至大肠杆菌DH5α进行克隆，然后经Sanger测序得到验证。最后将pET23ereA转化到大肠杆菌BL21中，以产生大肠杆菌ereA。

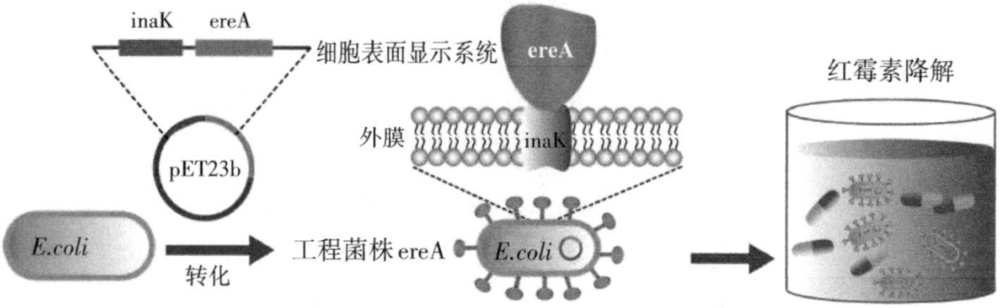

图4-7 大肠杆菌细胞表面显示的红霉素酯酶用于降解红霉素的示意图[136]

重组工程大肠杆菌ereA能够在盲肠、结肠等动物大肠中定植，并且其表面展示的红霉素酯酶表现出较高稳定性，pH为7时其活性最高。大肠杆菌ereA可在24 h内消除50 mg/L的红霉素，去除率达到100%。另外，通过测定红霉素及其降解产物的毒性，发现其降解产物对费氏弧菌产生的毒性很小。将收集的粪便在模拟堆肥温度下处理，发现仅通过堆肥高温处理难以降解红霉素，但大肠杆菌ereA的加入可以促进其降解。另外，在65 °C的堆肥温度下，工程细菌的*ereA*基因就会被消除，这避免了工程细菌释放到环境中，说明该工程菌具有环境安全性。

真菌漆酶是三种主要的氧化还原酶中的一种，由于其广泛的底物特异性，可以催化超过250种底物，并且在降解各种持久性有机污染物方面具有很高的活性[137，138]。例如，漆酶介质系统能够高效地去除磺胺类抗生素[139，140]。磺胺类抗生素多应用于动物繁殖及人类医学中，被人体、动物体排出后会残留在环境中[141]。此前研究发现地表水或污水废水处理厂中含有62.5%～94.8%的磺胺类抗生素及其残留物[142]。

基于这些背景信息，Li等在大肠杆菌（Nissle 1917）细胞表面表达真菌漆酶（图4-8）[143]。将工程化的生物催化剂作为饲料添加剂应用于白羽肉鸡的饲喂中，并评价其对白羽肉鸡体内肠道微生物平衡的影响，探究其能否提高肉鸡粪便中磺胺嘧啶的去除效率。

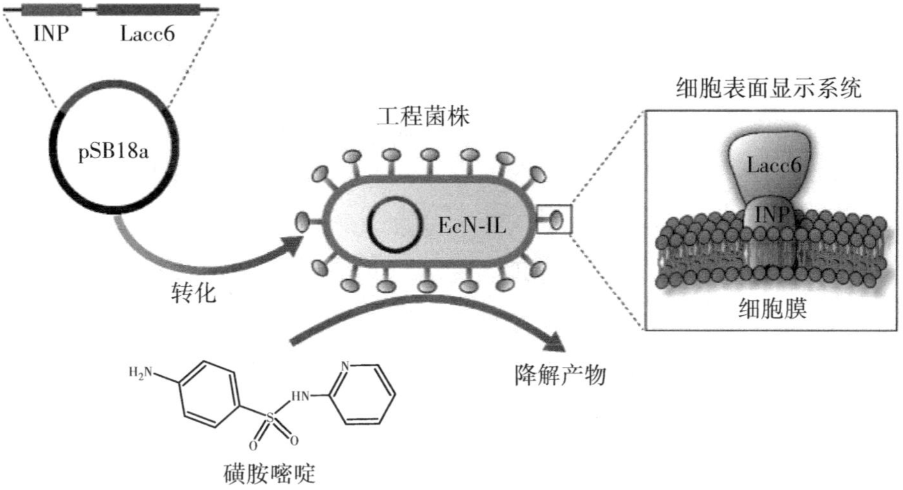

图4-8 大肠杆菌Nissle 1917细胞表面显示的用于降解磺胺嘧啶的漆酶示意图[143]

将GFP蛋白与INP-Lacc6融合，并转移到大肠杆菌Nissle 1917进行表达后，发现INP-Lacc6融合蛋白成功地锚定在EcN-IL菌株的表面。温度40 ℃和pH 5.0是EcN-IL菌株的最佳环境参数，并且发现该工程菌株能够在消化道定植。体外实验表明，工程菌株EcN-IL可有效加快磺胺嘧啶的氧化速率，提高其去除效率。体内实验显示，饲喂工程菌肉鸡粪便中磺胺嘧啶的残留量显著低于对照组，说明该工程菌可以在肉鸡肠道中定植并有效去除磺胺嘧啶。并且将EcN-IL添加至饲料中，可以减少长期磺胺嘧啶干预引起的门水平干扰，保持肉鸡肠道微生物群落的稳定状态。

将以上的两项研究与传统方法相比，可以看出这种新方法性价比高、更环保，为直接从家禽中去除抗生素残留提供了理论依据。同时也说明使用生物工程菌株减少抗生素在环境中的传播是可行的，并且通过构建表面工程细菌可能为生物修复带来巨大的前景[144]。此外，通过用感兴趣的蛋白质替换细胞表面展示酶，就可以获得具有不同功能的新型工程细菌，但是关于工程细菌在今后工作中的应用还面临一些挑战，对于该技术的安全性需要进一步检查和分析[144，145]。

4.3.3 利用肠道菌群治疗疾病的未来展望

肠道菌群是一个敏感且充满活力的生态系统，对于宿主的新陈代谢有非常重要的影响，并参与肥胖、糖尿病、心血管疾病、过敏、肠道炎症性疾病、癌症甚至精神疾病的病理生理学，可以通过生态失调影响疾病并介导各种疾病疗法的有效性[146]。越来越多的证据表明，利用肠道微生物群治疗可能会成为治疗某些疾病的新方法。

目前相关研究大多想通过调节肠道菌群组成进行疾病治疗，治疗策略包括饮食调节、使用益生元或益生菌补充剂和粪便微生物群移植。通过对高尿酸血症、高血压、结直肠癌及胃肠道癌等疾病的探究，发现基于肠道菌群的新型疗法有效并且具有巨大的潜力，为相关疾病的诊断、预防和治疗的临床应用提供了新的前景[147-149]。随着肠道菌群与这些疾病的关联信息日益丰富，肠道微生物群会成为开发治疗靶点的沃土[150，151]。研究人员预计，在不久的将来肠道菌群将作为疾病诊断的生物标志物，并且可通过“理想饮食”调节肠道菌群，从而治疗复杂疾病和达到代谢稳态[152]。

但是在今后研究中，还需要不断考虑并回答很多问题。例如，应该在疾病早期或晚期考虑肠道菌群调节，还是两者兼而有之？所有基于肠道菌群的治疗方法在人体是否具有有效性和安全性？能否发展改进对于肠道菌群操控的技术？综上所述，利用肠道菌群治疗疾病是一个非常重要且有趣的问题，也将会在进一步研究中被彻底探索。

4.4 肠道修复与病毒预防

病毒作为没有细胞结构的病原体，在细胞内寄生和自我复制的方式对人类健康构成了严重威胁。许多病毒，例如肠道病毒、细小病毒、人类免疫缺陷病毒、禽流感病毒等，在侵入人体后会影响肠道菌群的丰度和多样性[153]。因此，研究病毒与肠道菌群的关系将有助于治疗相关疾病，为临床诊疗奠定基础。

4.4.1 SARS-CoV-2与新冠感染患者的肠道菌群

严重急性呼吸综合征冠状病毒2（severe acute respiratory syndrome coronavirus 2，SARS-CoV-2）会导致严重的呼吸道疾病，被称为新型冠状病毒肺炎（coronavirus disease 2019，COVID–19）[154]。截至2020年8月12日，全球确诊的COVID-19病例超过2000万例，全球相关死亡人数超过74万例[155]。COVID-19不仅影响呼吸系统和心血管系统，还影响中枢神经系统和胃肠道系统[156]。

（1）肠道菌群反映新冠感染患者的疾病严重程度和免疫反应功能障碍

COVID-19患者经常出现腹泻和其他胃肠道症状，在武汉三家医院就诊的204例COVID-19患者，许多都有消化系统症状如腹泻，并且从发病到入院持续的时间很长[157，158]。有消化系统症状的患者从症状出现到入院的平均时间为9天，而有呼吸道症状的患者从症状出现到入院的平均时间为7.3天，并且有消化系统症状的患者表现出厌食、腹泻、呕吐或腹痛等临床症状[159]。

当病毒入侵后，会攻击肠道屏障，所以在肠道屏障完整性受损和肠道微生物群失衡时，会激活先天性和适应性免疫细胞将促炎细胞因子释放到循环系统中，导致全身炎症。一些肠道的信号通路可以通过树突状细胞进行炎症的调节（图4–9）。根据上述过程，通过激活上皮受体对先天宿主免疫进行免疫调节可能是在感染早期消除SARS-CoV-2新的治疗靶点[160]。

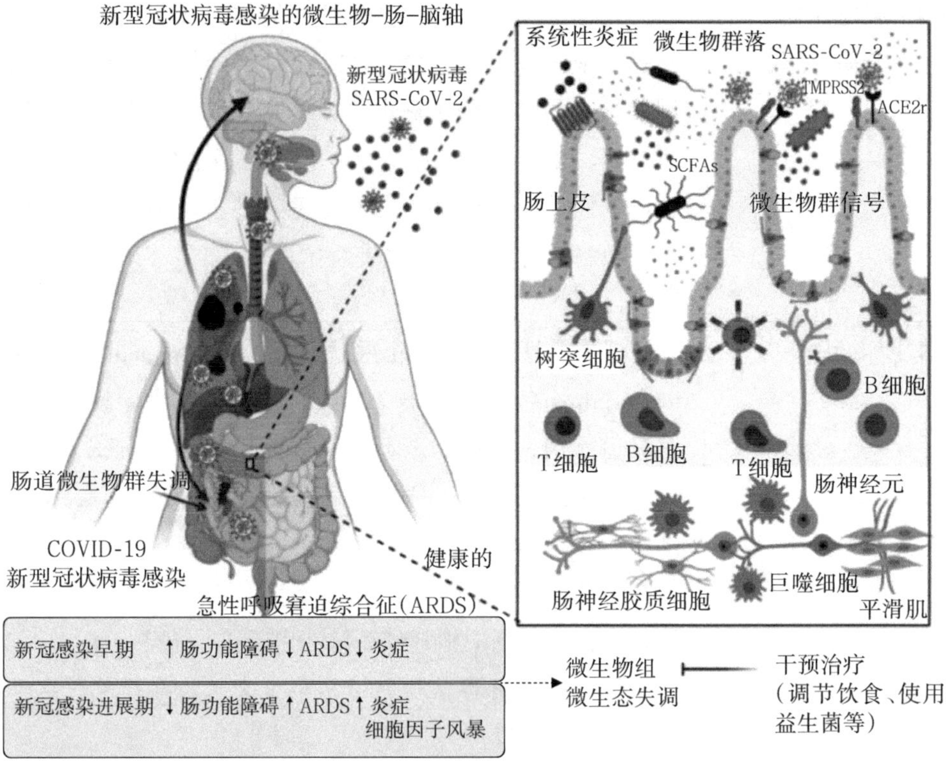

图4-9 SARS-CoV-2感染的说明性模型及其与肺-肠-脑轴和微生物组生态失调的关系[159]

从临床表现上看，腹泻的发生可能基于以下情况。在呼吸道症状出现后，病毒从肺部入侵肠道产生肠道损伤，机体释放炎症细胞，炎症细胞（包括中性粒细胞和淋巴细胞）进入肠黏膜，从而破坏肠道微生物群，致使发生腹泻反应。正如临床病例报告的那样，SARS-CoV-2可能对胃肠道系统有致病作用，会导致肠道微生物生态失调。另外，能够从粪便中分离出SARS-CoV-2，这表明粪便具有潜在的传染性，在疫情防控工作中更需要注意这一点[161, 162]。

（2）COVID-19潜在的预防和治疗策略

基于益生菌治疗高尿酸血症、心血管疾病及一些癌症的研究成果，可以将使用益生菌作为COVID-19潜在的预防和治疗策略。研究表明益生菌的抗病毒作用大多与乳酸杆菌和双歧杆菌有关[163]。如果将能增加益生菌能量来源的食物与益生菌混合或者将两种益生菌混合用于治疗疾病，可能会有不错的抗病毒效果。例如，通过将米糠与益生菌混合喂给无菌猪，两者的多种代谢产物能够改善肠道屏障功能，调节免疫反应，从而预防轮状病毒引起的腹泻[164]。一项临床试验中，让57例轮状病毒腹泻患者口服益生菌——长双歧杆菌BORI和嗜酸乳杆菌AD031，或者通过安慰剂配合标准腹泻治疗。结果显示，食用益生菌组患者病程明显短于对照组，能够更快地恢复，表明这两种益生菌的混合使用对轮状病毒有一定的抑制作用，并且与常规治疗方式相比更加高效[165]。

通过调节肠道菌群预防和治疗COVID-19，有以下几个潜在的策略：

第一，通过服用益生菌预防、治疗COVID-19。研究具体哪些益生菌能够有效保护感染SARS-CoV-2的患者，并且探究不同益生菌的组合对于感染后的修复是否有显著差异，并从中筛选出最优益生菌组合。

第二，可以考虑粪菌移植（fecal microbiota transplantation，FMT），这是一种新兴的治疗方法，并且对病毒性疾病也有一定的作用。研究表明，通过粪便移植可以使免疫缺陷小鼠产生抗病毒能力[166]。粪菌移植法可能更适用于免疫缺陷人群。

第三，肠道菌群作为人体的共生菌群，与人体的日常饮食息息相关，不同的饮食也会对肠道菌群产生不同的影响。研究表明，高纤维饮食中的可发酵碳水化合物对肠道屏障有积极影响，而西式饮食对肠道屏障有负面影响[167]。因此可以考虑饮食疗法，从而避免肠道菌群的病毒感染。

4.4.2 其他病毒与宿主的肠道菌群

研究发现，病毒感染与肠道微生物群能够相互作用，对肠道菌群产生显著影响，使肠道中有益细菌的数量减少，有害细菌的数量增加[168]。例如，呼吸道流感病毒感染后，肠道菌群会出现紊乱。其中肠杆菌的比例显著增加，而分段丝状菌和乳酸菌的比例显著降低，这是因为病毒通过诱导肠道菌群所介导的Th17细胞依赖性炎症，最终造成肠道免疫损伤[169]。另一项研究对于20例患有严重或复杂急性病毒性胃肠炎（acute gastroenteritis，AGE）的住院儿童和20名健康儿童的粪便进行了16S rRNA测序，研究发现，与正常对照组相比，复杂AGE患者肠道中弯曲杆菌科（Campylobacteraceae）、奈瑟菌科（Neisseriaceae）、甲基杆菌科（Methylobacteriaceae）、鞘磷脂单胞菌科（Sphingomonadaceae）、肠杆菌科（Enterobacteriaceae）的丰度增加，肠道微生物多样性显著降低[170]。

除了直接引起肠道菌群紊乱外，病毒还可以通过其他器官导致肠道菌群紊乱，目前许多研究集中于对脑-肠轴进行探索。据报道，微生物群在大脑功能调节中起着关键作用，肠道菌群与大脑之间的交流可能是双向的，即微生物群-肠道-脑轴[171]。当在小鼠颅内接种Theiler小鼠脑脊髓炎病毒时，拟普雷沃氏菌属（*Alloprevotella*）在急性期（14 dpi）相对丰度显著降低，而梭菌属XIVa（*Clostridium* XIVa）在28 dpi时丰度增加[172]。研究结果不仅表明病毒入侵大脑后，脑-肠轴的联系会导致肠道菌群失衡，而且还反映了肠道菌群在疾病的每个阶段都会不断变化。

综上所述，在研究疾病与肠道菌群关系时，不仅要深入探究具体的作用机制，而且还要对处在疾病不同阶段的肠道菌群进一步分析。也需要继续研究益生菌与肠道菌群以及病毒之间的关系，从而为今后实施针对性的诊疗提供理论基础。

（姜玉超　范竞文　韩华雯*）

参考文献

[1] ODEYEMI O A. Public health implications of microbial food safety and foodborne diseases in developing countries[J]. Food & nutrition research, 2016, 60: 1-2.

[2] 李泰然. 中国食源性疾病现状及管理建议[J]. 中华流行病学杂志, 2003, 24(8): 7-9.

[3] CHEN J, YING G G, DENG W J. Antibiotic residues in food: extraction, analysis, and human health concerns[J]. Journal of agricultural and food chemistry, 2019, 67(27): 7569-7586.

[4] BHATTACHARYYA S, DAS C. Foodborne infections and food safety[J]. Eastern journal of medical sciences, 2022, 7(3): 60-63.

[5] 中华预防医学会微生态学分会. 中国消化道微生态调节剂临床应用专家共识(2016版)[J]. 中国实用内科杂志, 2016, 36(10): 858-869.

[6] 李锋. 急性食源性胃肠炎的预防与治疗研究[J]. 大众科技, 2018, 20(8): 90-92.

[7] QUIGLEY E M. Prebiotics and probiotics; modifying and mining the microbiota[J]. Pharmacological research, 2010, 61(3): 213-218.

[8] 方海明, 王佳佳, 章礼久. 酪酸杆菌双歧杆菌二联活菌联合马来酸曲美布汀治疗急性胃肠炎疗效观察[J]. 安徽医药, 2013, 17(3): 1207-1209.

[9] 熊祖明, 袁杰利. 酪酸梭菌的研究与应用进展[J]. 中国微生态学杂志, 2011, 23(12): 1143-1145.

[10] BRENNAN A M. Development of synthetic biotics as treatment for human diseases[J]. Synthetic biology, 2022, 7(1): 1-7.

[11] YANG D F, LIU Y L, LIU S, et al. Exposure to heavy metals and its association with DNA oxidative damage in municipal waste incinerator workers in Shenzhen, China[J]. Chemosphere, 2020, 250: 1-9.

[12] NOUHA K, KUMAR R S, TYAGI R D. Heavy metals removal from wastewater using extracellular polymeric substances produced by *Cloacibacterium normanense* in wastewater sludge supplemented with crude glycerol and study of extracellular polymeric substances extraction by different methods [J]. Bioresource technology, 2016,

212:120-129.

[13] MADANI F Z, HAFIDA M, MERZOUK S A, et al. Hemostatic, inflammatory, and oxidative markers in pesticide user farmers[J]. Biomarkers, 2016, 21(2): 138-145.

[14] NARAYANA K. An aminoglycoside antibiotic gentamycin induces oxidative stress, reduces antioxidant reserve and impairs spermatogenesis in rats[J]. Journal of toxicological sciences, 2008, 33(1): 85-96.

[15] ERCAL N, GURERORHAN H, AYKINBURNS N. Toxic metals and oxidative stress part I: Mechanisms involved in metal-induced oxidative damage[J]. Current topics in medicinal chemistry, 2001, 1(6): 529-539.

[16] REHMAN K, FATIMA F, WAHEED I, et al. Prevalence of exposure of heavy metals and their impact on health consequences[J]. Journal of cellular biochemistry, 2018, 119(1): 157-184.

[17] POET T S, WU H, KOUSBA A A, et al. In vitro rat hepatic and intestinal metabolism of the organophosphate pesticides chlorpyrifos and diazinon[J]. Toxicological sciences, 2003, 72(2): 193-200.

[18] 王赛怡, 王逸君, 赵亚洲. 土壤重金属污染及其植物修复研究进展[J]. 农学学报, 2023, 13(2): 20-23.

[19] WU G F, XIAO X P, FENG P Y, et al. Gut remediation: a potential approach to reducing chromium accumulation using *Lactobacillus plantarum* TW1-1[J]. Scientific reports, 2017, 7(1): 1-12.

[20] OJUEDERIE O B, BABALOLA O O. Microbial and plant-assisted bioremediation of heavy metal polluted environments: a review[J]. International journal of environmental research and public health, 2017, 14(12): 1-26.

[21] MONACHESE M, BURTON J P, REID G. Bioremediation and tolerance of humans to heavy metals through microbial processes: a potential role for probiotics[J]. Applied and environmental microbiology, 2012, 78(18): 6397-6404.

[22] ZHAI Q X, WANG G, ZHAO J X, et al. Protective effects of *Lactobacillus plantarum* CCFM8610 against chronic cadmium toxicity in mice indicate routes of protection besides intestinal sequestration[J]. Applied and environmental microbiology, 2014, 80(13): 4063-4071.

[23] TIAN F W, ZHAI Q X, ZHAO J X, et al. *Lactobacillus plantarum* CCFM8661 alleviates lead toxicity in mice[J]. Biological trace element research, 2012, 150(1): 264–271.

[24] TIAN F W, XIAO Y, LI X X, et al. Protective effects of *Lactobacillus plantarum* CCFM8246 against copper toxicity in mice[J]. PLoS one, 2015, 10(11): 1–16.

[25] YU L L, ZHAI Q X, LIU X M, et al. *Lactobacillus plantarum* CCFM639 alleviates aluminium toxicity[J]. Applied microbiology and biotechnology, 2016, 100(4): 1891–1900.

[26] GHOSH S, PRAMANIK S. Structural diversity, functional aspects and future therapeutic applications of human gut microbiome[J]. Archives of microbiology, 2021, 203(9): 5281–5308.

[27] BRETON J, DANIEL C, DEWULF J, et al. Gut microbiota limits heavy metals burden caused by chronic oral exposure[J]. Toxicology letters, 2013, 222(2): 132–138.

[28] GANESH B P, FULTZ R, AYYASWAMY S, et al. Microbial interactions with the intestinal epithelium and beyond: focusing on immune cell maturation and homeostasis[J]. Current pathobiology reports, 2018, 6(1): 47–54.

[29] XING S C, HUANG C B, MI J D, et al. *Bacillus coagulans* R11 maintained intestinal villus health and decreased intestinal injury in lead-exposed mice by regulating the intestinal microbiota and influenced the function of faecal microRNAs[J]. Environmental pollution, 2019, 255(2): 113–139.

[30] SWANSON K S, GIBSON G R, HUTKINS R, et al. The international scientific association for probiotics and prebiotics (ISAPP) consensus statement on the definition and scope of synbiotics[J]. Nature reviews gastroenterology & hepatology, 2020, 17(11): 687–701.

[31] RAD A H, ABBASALIZADEH S, VAZIFEKHAH S, et al. The future of diabetes management by healthy probiotic microorganisms[J]. Current diabetes reviews, 2017, 13(6): 582–589.

[32] SARFRAZ F, FAROOQ U, SHAFI A, et al. Hypolipidaemic effects of synbiotic yoghurt in rabbits[J]. International journal of dairy technology, 2019, 72(4): 545–550.

[33] LEE C S, LEE S H, KIM S H. Bone-protective effects of *Lactobacillus plantarum* B719-fermented milk product[J]. International journal of dairy technology, 2020, 73(4): 706-717.

[34] ANSARI F, POURJAFAR H, TABRIZI A, et al. The effects of probiotics and prebiotics on mental disorders: a review on depression, anxiety, alzheimer, and autism spectrum disorders[J]. Current pharmaceutical biotechnology, 2020, 21(7): 555-565.

[35] ROOBAB U, BATOOL Z, MANZOOR M F, et al. Sources, formulations, advanced delivery and health benefits of probiotics[J]. Current opinion in food science, 2020, 32: 17-28.

[36] FENG P Y, YE Z, KAKADE A, et al. A review on gut remediation of selected environmental contaminants: possible roles of probiotics and gut microbiota[J]. Nutrients, 2018, 11(1): 1-19.

[37] BISANZ J E, ENOS M K, MWANGA J R, et al. Randomized open-label pilot study of the influence of probiotics and the gut microbiome on toxic metal levels in tanzanian pregnant women and school children[J]. Journal of microbiology, 2014, 5(5): 1580-1594.

[38] ZHAI Q X, LIU Y, WANG C, et al. Increased cadmium excretion due to oral administration of *Lactobacillus plantarum* strains by regulating enterohepatic circulation in mice[J]. Journal of agricultural and food chemistry, 2019, 67(17): 3956-3965.

[39] BALLARD J W O, TOWARNICKI S G. Mitochondria, the gut microbiome and ROS[J]. Cellular signalling, 2020, 75: 1-11.

[40] BHATTACHARYYA A, CHATTOPADHYAY R, MITRA S, et al. Oxidative stres: an essential factor in the pathogenesis of gastrointestinal mucosal diseases[J]. Physiological reviews, 2014, 94(2): 329-354.

[41] CHI L, BIAN X M, GAO B, et al. The effects of an environmentally relevant level of arsenic on the gut microbiome and its functional metagenome[J]. Toxicological sciences, 2017, 160(2): 193-204.

[42] PEGHAIRE E, MONE A, DELBAC F, et al. A *Pediococcus* strain to rescue honeybees by decreasing *Nosema ceranae*- and pesticide-induced adverse effects[J]. Pesticide biochemistry and physiology, 2020, 163: 138-146.

[43] TRINDER M, BISANZ J E, BURTON J P, et al. Probiotic lactobacilli: a potential prophylactic treatment for reducing pesticide absorption in humans and wildlife [J]. Beneficial microbes, 2015, 6(6): 841–847.

[44] ZHANG D, LI C, SHI R R, et al. *Lactobacillus fermentum* JX306 restrain d-galactose-induced oxidative stress of mice through its antioxidant activity [J]. Polish journal of microbiology, 2020, 69(2): 205–215.

[45] LI F, HUANG G B, TAN F, et al. *Lactobacillus plantarum* KSFY06 on d-galactose-induced oxidation and aging in Kunming mice [J]. Food science & nutrition, 2020, 8(1): 379–389.

[46] ZHAO J C, TIAN F W, YAN S, et al. *Lactobacillus plantarum* CCFM10 alleviating oxidative stress and restoring the gut microbiota in d-galactose-induced aging mice [J]. Food & function, 2018, 9(2): 917–924.

[47] FENG P Y, YANG J F, ZHAO S, et al. Human supplementation with *Pediococcus acidilactici* GR-1 decreases heavy metals levels through modifying the gut microbiota and metabolome [J]. NPJ biofilms and microbiomes, 2022, 8(1): 1–18.

[48] LITVAK Y, BYNDLOSS M X, TSOLIS R M, et al. Dysbiotic *Proteobacteria* expansion: a microbial signature of epithelial dysfunction [J]. Current opinion in microbiology, 2017, 39: 1–6.

[49] LI J, LEI R H, LI X, et al. The antihyperlipidemic effects of fullerenol nanoparticles via adjusting the gut microbiota in vivo [J]. Particle and fibre toxicology, 2018, 15(1): 1–11.

[50] CECCARANI C, BASSANINI G, MONTANARI C, et al. Proteobacteria overgrowth and butyrate-producing taxa depletion in the gut microbiota of glycogen storage disease type 1 patients [J]. Metabolites, 2020, 10(4): 1–16.

[51] TEEM H, SEPPO S, JUSSI M, et al. Reversible surface binding of cadmium and lead by lactic acid and bifidobacteria [J]. International journal of food microbiology, 2008, 125(2): 170–175.

[52] TOPCU A, BULAT T. Removal of cadmium and lead from aqueous solution by *Enterococcus faecium* strains [J]. Journal of food science, 2010, 75(1): 13–17.

[53] ZHAI Q X, WANG J, CEN S, et al. Modulation of the gut microbiota by a galactooligosaccharide protects against heavy metal lead accumulation in mice [J]. Food

& function,2019,10(6):3768–3781.

[54] ZHU J M, YU L L, SHEN X D, et al. Protective effects of *Lactobacillus plantarum* CCFM8610 against acute toxicity caused by different food-derived forms of cadmium in mice[J]. International journal of molecular sciences, 2021, 22(20): 1–15.

[55] RAD A H, AKBARZADEH F, MEHRABANY E V. Which are more important: prebiotics or probiotics[J]. Nutrition, 2012, 28(11): 1196–1197.

[56] HUI C Y, GUO Y, ZHANG W, et al. Surface display of PbrR on *Escherichia coli* and evaluation of the bioavailability of lead associated with engineered cells in mice [J]. Scientific reports, 2018, 8(1): 1–12.

[57] VAN DER KUIJP T J, HUANG L, CHERRY C R. Health hazards of China's lead-acid battery industry: a review of its market drivers, production processes, and health impacts[J]. Environmental health, 2013, 12: 1–10.

[58] TAYLOR M P, FORBES M K, OPESKIN B, et al. The relationship between atmospheric lead emissions and aggressive crime: an ecological study[J]. Environmental health, 2016, 15: 1–10.

[59] PERUGINI M, VISCIANO P, MANERA M, et al. Heavy metal (As, Cd, Hg, Pb, Cu, Zn, Se) concentrations in muscle and bone of four commercial fish caught in the central Adriatic Sea, Italy [J]. Environmental monitoring and assessment, 2014, 186 (4): 2205–2213.

[60] YI Y J, YANG Z F, ZHANG S H. Ecological risk assessment of heavy metals in sediment and human health risk assessment of heavy metals in fishes in the middle and lower reaches of the Yangtze River basin [J]. Environmental pollution, 2011, 159 (10): 2575–2585.

[61] PLEGARIA J S, DZUL S P, ZUIDERWEG E R P, et al. Apoprotein structure and metal binding characterization of a *de novo* designed peptide, α_3DIV, that sequesters toxic heavy metals[J]. Biochemistry, 2015, 54(18): 2858–2873.

[62] XIAO S, GAO W, CHEN Q F, et al. Overexpression of membrane-associated acyl-CoA-binding protein ACBP1 enhances lead tolerance in arabidopsis [J]. Plant journal, 2008, 54(1): 141–151.

[63] LI Q Q, YAN Q, CHEN J S, et al. Molecular characterization of an ice nucleation protein variant (inaq) from pseudomonas syringae and the analysis of its trans-

membrane transport activity in *Escherichia coli*[J]. International journal of biological sciences, 2012, 8(8): 1097–1108.

[64] CHEN X, YANG J F, LING Z M, et al. Gut *Escherichia coli* expressing Pb^{2+}-adsorption protein reduces lead accumulation in grass carp, *Ctenopharyngodon idellus* [J]. Environmental pollution, 2021, 276: 1–9.

[65] LI T T, LONG M, GATESOUPE F J, et al. Comparative analysis of the intestinal bacterial communities in different species of carp by pyrosequencing[J]. Microbial ecology, 2015, 69(1): 25–36.

[66] XIE S L, ZHOU A G, WEI T L, et al. Nanoplastics induce more serious microbiota dysbiosis and inflammation in the gut of adult zebrafish than microplastics[J]. Bulletin of environmental contamination and toxicology, 2021, 107(4): 640–650.

[67] XIA J Z, JIN C Y, PAN Z H, et al. Chronic exposure to low concentrations of lead induces metabolic disorder and dysbiosis of the gut microbiota in mice[J]. Science of the total environment, 2018, 631: 439–448.

[68] JIN C Y, LUO T, ZHU Z H, et al. Imazalil exposure induces gut microbiota dysbiosis and hepatic metabolism disorder in zebrafish[J]. Comparative biochemistry and physiology C-toxicology & pharmacology, 2017, 202: 85–93.

[69] PAUL N, CHAKRABORTY S, SENGUPTA M. Lead toxicity on non-specific immune mechanisms of freshwater fish *Channa punctatus*[J]. Aquatic toxicology, 2014, 152: 105–112.

[70] LI B Y, ZHANG Y M, MA D X, et al. Mercury nano-trap for effective and efficient removal of mercury (Ⅱ) from aqueous solution [J]. Nature communications, 2014, 5: 1–7.

[71] YU J G, YUE B Y, WU X W, et al. Removal of mercury by adsorption: a review[J]. Environmental science and pollution research, 2016, 23(6): 5056–5076.

[72] LIU Y R, LU X, ZHAO L D, et al. Effects of cellular sorption on mercury bioavailability and methylmercury production by *Desulfovibrio desulfuricans* ND132 [J]. Environmental science & technology, 2016, 50(24): 13335–13341.

[73] ZHANG W, ZHANG X, TIAN Y L, et al. Risk assessment of total mercury and methylmercury in aquatic products from offshore farms in China[J]. Journal of hazardous materials, 2018, 354: 198–205.

[74] CERVENY D, TUREK J, GRABIC R, et al. Young-of-the-year fish as a prospective bioindicator for aquatic environmental contamination monitoring[J]. Water research, 2016, 103:334–342.

[75] BECKERS F, RINKLEBE J. Cycling of mercury in the environment: sources, fate, and human health implications: a review[J]. Critical reviews in environmental science and technology, 2017, 47(9):693–794.

[76] TUZEN M, KARAMAN I, CITAK D, et al. Mercury(Ⅱ) and methyl mercury determinations in water and fish samples by using solid phase extraction and cold vapour atomic absorption spectrometry combination[J]. Food and chemical toxicology, 2009, 47(7):1648–1652.

[77] WANG H S, XU W F, CHEN Z J, et al. In vitro estimation of exposure of Hong Kong residents to mercury and methylmercury via consumption of market fishes [J]. Journal of hazardous materials, 2013, 248:387–393.

[78] LIU M R, LU X, KHAN A, et al. Reducing methylmercury accumulation in fish using *Escherichia coli* with surface-displayed methylmercury-binding peptides[J]. Journal of hazardous materials, 2019, 367:35–42.

[79] WANG P, WU J, LIU L X, et al. A peptide-based fluorescent chemosensor for measuring cadmium ions in aqueous solutions and live cells[J]. Dalton transactions, 2015, 44(41):18057–18064.

[80] JUNG H C, LEBEAULT J M, PAN J G. Surface display of *Zymomonas mobilis* levansucrase by using the ice-nucleation protein of *Pseudomonas syringae*[J]. Nature biotechnology, 1998, 16(6):576–580.

[81] POLAK-JUSZCZAK L. Distribution of organic and inorganic mercury in the tissues and organs of fish from the southern Baltic Sea[J]. Environmental science and pollution research, 2018, 25(34):34181–34189.

[82] ZHUANG X J, XIONG L S, LI L, et al. Alterations of gut microbiota in patients with irritable bowel syndrome: a systematic review and meta-analysis[J]. Journal of gastroenterology and hepatology, 2017, 32(1):28–38.

[83] SHAO D D, KANG Y, WU S C, et al. Effects of sulfate reducing bacteria and sulfate concentrations on mercury methylation in freshwater sediments[J]. Science of the total environment, 2012, 424:331–336.

[84] LIU M R, KAKADE A, LIU P, et al. Hg^{2+}-binding peptide decreases mercury ion accumulation in fish through a cell surface display system[J]. Science of the total environment, 2019, 659: 540–547.

[85] COMBER S, GARDNER M. Chromium redox speciation in natural waters [J]. Journal of environmental monitoring, 2003, 5(3): 410–413.

[86] SUN Z L, KONG J, KONG W T. Characterization of a cryptic plasmid pD403 from *Lactobacillus plantarum* and construction of shuttle vectors based on its replicon [J]. Molecular biotechnology, 2010, 45(1): 24–33.

[87] FENG P Y, YE Z, HAN H W, et al. Tibet plateau probiotic mitigates chromate toxicity in mice by alleviating oxidative stress in gut microbiota[J]. Communications biology, 2020, 3(1): 1–12.

[88] CARPENTER D O. Health effects of persistent organic pollutants: the challenge for the Pacific Basin and for the world [J]. Reviews on environmental health, 2011, 26(1): 61–69.

[89] 马卓,龚书识,苏林冲,等.痛风现状及其患者依从性情况[J].世界最新医学信息文摘,2018,18(45):99–100.

[90] LIU R, HAN C, WU D, et al. Prevalence of hyperuricemia and gout in mainland china from 2000 to 2014: a systematic review and meta-analysis [J]. Biomed research international, 2015(2015): 1–12.

[91] 谢丽玲,贺盼攀,秦献辉,等.高尿酸血症治疗的研究进展[J].生物医学转化,2021,2(4):34–40.

[92] 柳艳云,杨亚强,段学辉.西北传统美食浆水的研究进展[J]食品与发酵工业,2017,43(11):262–267.

[93]张铁,王玉丽,陈晓前,等.传统发酵食品——浆水中微生物的分离与初步鉴定[J].食品科学,2007(1):219–222.

[94]李雪萍,李建宏,李敏权,等.浆水中降胆固醇乳酸菌的筛选及其功能特性[J].微生物学报,2015,55(8):1001–1009.

[95]WU Y, YE Z, FENG P, et al. *Limosilactobacillus fermentum* JL-3 isolated from "Jiangshui" ameliorates hyperuricemia by degrading uric acid [J]. Gut microbes, 2021, 13(1): 1–18.

[96]ZHAO S, FENG P, HU X, et al. Probiotic *Limosilactobacillus fermentum* GR-

3 ameliorates human hyperuricemia via degrading and promoting excretion of uric acid [J]. iScience,2022,25(10):1–18.

[97]WANG J,CHEN G,CHRISTIE P,et al. Occurrence and risk assessment of phthalate esters (PAEs) in vegetables and soils of suburban plastic film greenhouses [J]. Science of the total environment,2015,523:129–137.

[98]KAY V R,BLOOM M S,FOSTER W G. Reproductive and developmental effects of phthalate diesters in males [J]. Critical reviews in toxicology,2014,44(6):467–498.

[99]ZHANG L D,LI H C,CHONG T,et al. Prepubertal exposure to genistein alleviates di-(2-ethylhexyl)phthalate induced testicular oxidative stress in adult rats [J]. Biomed research international,2014,2014:1–9.

[100]HU J,RAIKHEL V,GOPALAKRISHNAN K,et al. Effect of postnatal low-dose exposure to environmental chemicals on the gut microbiome in a rodent model [J]. Microbiome,2016,4(1):1–11.

[101]AL-ASMAKH M,STUKENBORG J B,REDA A,et al. The gut microbiota and developmental programming of the testis in mice [J]. PLoS one,2014,9(8):1–10.

[102]ZHAI Q,WANG G,ZHAO J,et al. Protective effects of *Lactobacillus plantarum* CCFM8610 against chronic cadmium toxicity in mice indicate routes of protection besides intestinal sequestration [J]. Applied and environmental microbiology,2014,80(13):4063–4071.

[103]POUTAHIDIS T,SPRINGER A,LEVKOVICH T,et al. Probiotic microbes sustain youthful serum testosterone levels and testicular size in aging mice [J]. PLoS one,2014,9(1):1–12.

[104]TIAN X,YU Z,FENG P,et al. *Lactobacillus plantarum* TW1-1 alleviates diethylhexylphthalate-induced testicular damage in mice by modulating gut microbiota and decreasing inflammation [J]. Frontiers in cellular and infection microbiology,2019,9:1–16.

[105]REHM J,SAMOKHVALOV A V,SHIELD K D. Global burden of alcoholic liver diseases [J]. Journal of hepatology,2013,59(1):160–168.

[106]CRABB D W,MATSUMOTO M,CHANG D,et al. Overview of the role of alcohol dehydrogenase and aldehyde dehydrogenase and their variants in the genesis of al-

cohol-related pathology [J]. Proceedings of the nutrition society, 2004, 63(1): 49–63.

[107] BASUROY S, SHETH P, MANSBACH C M, et al. Acetaldehyde disrupts tight junctions and adherens junctions in human colonic mucosa: protection by EGF and l-glutamine [J]. American journal of physiology-gastrointestinal and liver physiology, 2005, 289(2): 367–375.

[108] LEUNG T M, NIETO N. CYP2E1 and oxidant stress in alcoholic and non-alcoholic fatty liver disease [J]. Journal of hepatology, 2013, 58(2): 395–398.

[109] SCHNABL B, BRENNER D A. Interactions between the intestinal microbiome and liver diseases [J]. Gastroenterology, 2014, 146(6): 1513–1524.

[110] MINEMURA M, SHIMIZU Y. Gut microbiota and liver diseases [J]. World journal of gastroenterology, 2015, 21(6): 1691–1702.

[111] CHEN P, STÄRKEL P, TURNER J R, et al. Dysbiosis-induced intestinal inflammation activates tumor necrosis factor receptor I and mediates alcoholic liver disease in mice [J]. Hepatology, 2015, 61(3): 883–894.

[112] TUOMISTO S, PESSI T, COLLIN P, et al. Changes in gut bacterial populations and their translocation into liver and ascites in alcoholic liver cirrhotics [J]. BMC gastroenterology, 2014, 14(1): 1–8.

[113] HAO L, SUN Q, ZHONG W, et al. Mitochondria-targeted ubiquinone (MitoQ) enhances acetaldehyde clearance by reversing alcohol-induced posttranslational modification of aldehyde dehydrogenase 2: a molecular mechanism of protection against alcoholic liver disease [J]. Redox biology, 2018, 14: 626–636.

[114] KIM W G, KIM H I, KWON E K, et al. *Lactobacillus plantarum* LC27 and *Bifidobacterium longum* LC67 mitigate alcoholic steatosis in mice by inhibiting LPS-mediated NF-κB activation through restoration of the disturbed gut microbiota [J]. Food & function, 2018, 9(8): 4255–4265.

[115] VAN BEEK A A, HOOGERLAND J A, BELZER C, et al. Interaction of mouse splenocytes and macrophages with bacterial strains in vitro: the effect of age in the immune response [J]. Beneficial microbes, 2016, 7(2): 275–287.

[116] TIAN X, LI R, JIANG Y, et al. *Bifidobacterium breve* ATCC15700 pretreatment prevents alcoholic liver disease through modulating gut microbiota in mice exposed to chronic alcohol intake [J]. Journal of functional foods, 2020, 72: 1–12.

[117] KAFAEI R, PAPARI F, SEYEDABADI M, et al. Occurrence, distribution, and potential sources of antibiotics pollution in the water - sediment of the northern coastline of the Persian Gulf, Iran [J]. Science of the total environment, 2018, 627: 703–712.

[118] LI B, ZHANG T. Different removal behaviours of multiple trace antibiotics in municipal wastewater chlorination [J]. Water research, 2013, 47(9): 2970–2982.

[119] FENG L, CASAS M E, OTTOSEN L D M, et al. Removal of antibiotics during the anaerobic digestion of pig manure [J]. Science of the total environment, 2017, 603: 219–225.

[120] TRAN N H, CHEN H, REINHARD M, et al. Occurrence and removal of multiple classes of antibiotics and antimicrobial agents in biological wastewater treatment processes [J]. Water research, 2016, 104: 461–472.

[121] IVANOV A E, HALTHUR T, LJUNGGREN L. Flow permeable composites of lignin and poly (vinyl alcohol): towards removal of bisphenol A and erythromycin from water [J]. Journal of environmental chemical engineering, 2016, 4 (2): 1432–1441.

[122] MICHAEL I, RIZZO L, MCARDELL C S, et al. Urban wastewater treatment plants as hotspots for the release of antibiotics in the environment: a review [J]. Water research, 2013, 47(3): 957–995.

[123] HUANG Y, DONG H, SHANG B, et al. Characterization of animal manure and cornstalk ashes as affected by incineration temperature [J]. Applied energy, 2011, 88(3): 947–952.

[124] SKOULOU V, ZABANIOTOU A. Investigation of agricultural and animal wastes in Greece and their allocation to potential application for energy production [J]. Renewable and sustainable energy reviews, 2007, 11(8): 1698–1719.

[125] GUO A, GU J, WANG X, et al. Effects of superabsorbent polymers on the abundances of antibiotic resistance genes, mobile genetic elements, and the bacterial community during swine manure composting [J]. Bioresource technology, 2017, 244: 658–663.

[126] YIN F, DONG H, JI C, et al. Effects of anaerobic digestion on chlortetracycline and oxytetracycline degradation efficiency for swine manure [J]. Waste manage-

ment,2016,56:540-546.

[127]EZZARIAI A,HAFIDI M,KHADRA A,et al. Human and veterinary antibiotics during composting of sludge or manure: global perspectives on persistence, degradation, and resistance genes [J]. Journal of hazardous materials,2018,359:465-481.

[128]YANG X,YOU S,SHI Y,et al. The research about the removal of antibiotic pollutants in water pollution by adsorption materials and photocatalytic materials [J]. IOP conference series: earth and environmental science,2018,186(3):1-5.

[129]OU H,CHEN Q,PAN J,et al. Selective removal of erythromycin by magnetic imprinted polymers synthesized from chitosan - stabilized Pickering emulsion [J]. Journal of hazardous materials,2015,289:28-37.

[130]SINHA R K,BHARAMBE G,RYAN D. Converting wasteland into wonderland by earthworms-a low-cost nature's technology for soil remediation: a case study of vermiremediation of PAHs contaminated soil [J]. The environmentalist,2008,28(4):466-475.

[131]CAI M,MA S,HU R,et al. Systematic characterization and proposed pathway of tetracycline degradation in solid waste treatment by Hermetia illucens with intestinal microbiota [J]. Environmental pollution,2018,242:634-642.

[132]SMITH M R,KHERA E,WEN F. Engineering novel and improved biocatalysts by cell surface display [J]. Industrial & engineering chemistry research,2015,54(16):4021-4032.

[133]YANG T,CHEN M L,WANG J H. Genetic and chemical modification of cells for selective separation and analysis of heavy metals of biological or environmental significance [J]. TrAC trends in analytical chemistry,2015,66:90-102.

[134]DE CAZES M,BELLEVILLE M P,PETIT E,et al. Erythromycin degradation by esterase (EreB) in enzymatic membrane reactors [J]. Biochemical engineering journal,2016,114:70-78.

[135]PÉREZ T,SIRÉS I,BRILLAS E,et al. Solar photoelectro-Fenton flow plant modeling for the degradation of the antibiotic erythromycin in sulfate medium [J]. Electrochimica acta,2017,228:45-56.

[136]LIU M,FENG P,KAKADE A,et al. Reducing residual antibiotic levels in animal feces using intestinal *Escherichia coli* with surface-displayed erythromycin ester-

ase [J]. Journal of hazardous materials, 2020, 388: 1-11.

[137] CHEN Y, STEMPLE B, KUMAR M, et al. Cell surface display fungal laccase as a renewable biocatalyst for degradation of persistent micropollutants bisphenol a and sulfamethoxazole [J]. Environmental science & technology, 2016, 50(16): 8799-8808.

[138] BALDRIAN P.Fungal laccases - occurrence and properties [J].Fems microbiology reviews, 2006, 30(2): 215-242.

[139] WENG S S, KU K L, LAI H T. The implication of mediators for enhancement of laccase oxidation of sulfonamide antibiotics [J]. Bioresource technology, 2012, 113: 259-264.

[140] ZHUO R, YU H, YUAN P, et al. Heterologous expression and characterization of three laccases obtained from *Pleurotus ostreatus* HAUCC 162 for removal of environmental pollutants [J]. Journal of hazardous materials, 2018, 344: 499-510.

[141] VAN BOECKEL T P, BROWER C, GILBERT M, et al. Global trends in antimicrobial use in food animals [J]. Proceedings of the national academy of sciences, 2015, 112(18): 5649-5654.

[142] CHEN H, LIU S, XU X R, et al. Antibiotics in the coastal environment of the hailing bay region, south china sea: spatial distribution, source analysis and ecological risks [J]. Marine pollution bulletin, 2015, 95(1): 365-373.

[143] LI R, ZHOU T, KHAN A, et al. Feed-additive of bioengineering strain with surface-displayed laccase degrades sulfadiazine in broiler manure and maintains intestinal flora structure [J]. Journal of hazardous materials, 2021, 406: 1-9.

[144] SINGH J S, ABHILASH P C, SINGH H B, et al. Genetically engineered bacteria: an emerging tool for environmental remediation and future research perspectives [J]. Gene, 2011, 480(1): 1-9.

[145] HWANG I Y, KOH E, WONG A, et al. Engineered probiotic *Escherichia coli* can eliminate and prevent *Pseudomonas aeruginosa* gut infection in animal models [J]. Nature communications, 2017, 8(1): 1-11.

[146] RIAZ RAJOKA M S, MEHWISH H M, XIONG Y, et al. Gut microbiota targeted nanomedicine for cancer therapy: challenges and future considerations [J]. Trends in food science & technology, 2021, 107: 240-251.

[147]DRAPKINA O M, YAFAROVA A A, KABUROVA A N, et al. Targeting gut microbiota as a novel strategy for prevention and treatment of hypertension, atrial fibrillation and heart failure: current knowledge and future perspectives [J]. Biomedicines, 2022, 10(8): 1-17.

[148]LI J, ZHANG A H, WU F F, et al. Alterations in the gut microbiota and their metabolites in colorectal cancer: recent progress and future prospects [J]. Frontiers in oncology, 2022, 12: 1-14.

[149]KOUZU K, TSUJIMOTO H, KISHI Y, et al. Bacterial translocation in gastrointestinal cancers and cancer treatment[J]. Biomedicines. 2022, 10(2): 1-15.

[150]BORODY T J, WARREN E F, LEIS S M, et al. Bacteriotherapy using fecal flora: toying with human motions [J]. Journal of clinical gastroenterology, 2004, 38(6): 475-483.

[151]KINROSS J M, DARZI A W, NICHOLSON J K. Gut microbiome-host interactions in health and disease [J]. Genome medicine, 2011, 3(3): 1-12.

[152]NIE Y, LUO F, LIN Q. Dietary nutrition and gut microflora: a promising target for treating diseases [J]. Trends in food science & technology, 2018, 75: 72-80.

[153]LV Z, XIONG D, SHI J, et al. The interaction between viruses and intestinal microbiota: a review [J]. Current microbiology, 2021, 78(10): 3597-3608.

[154]CAO Y, CAI K, XIONG L. Coronavirus disease 2019: a new severe acute respiratory syndrome from wuhan in china [J]. Acta virologica, 2020, 64(2): 245-250.

[155]DONG E, DU H, GARDNER L. An interactive web-based dashboard to track COVID-19 in real time [J]. The lancet infectious diseases, 2020, 20(5): 533-534.

[156]BAIG A M, SANDERS E C. Potential neuroinvasive pathways of SARS-COV-2: deciphering the spectrum of neurological deficit seen in coronavirus disease-2019 (COVID-19)[J]. Journal of medical virology, 2020, 92(10): 1845-1857.

[157]JIN X, LIAN J S, HU J H, et al. Epidemiological, clinical and virological characteristics of 74 cases of coronavirus-infected disease 2019 (COVID-19) with gastrointestinal symptoms [J]. Gut, 2020, 69(6): 1-8.

[158]PAN L, MU M, YANG P, et al. Clinical characteristics of COVID-19 patients with digestive symptoms in hubei, china: a descriptive, cross-sectional, multi-

center study [J]. Official journal of the american college of gastroenterology, 2020, 115(5): 766–773.

[159] VILLAPOL S. Gastrointestinal symptoms associated with COVID-19: impact on the gut microbiome [J]. Translational research, 2020, 226: 57–69.

[160] GOLONKA R M, SAHA P, YEOH B S, et al. Harnessing innate immunity to eliminate SARS-CoV-2 and ameliorate COVID-19 disease [J]. Physiological genomics, 2020, 52(5): 217–221.

[161] MEINI S, ZINI C, PASSALEVA M T, et al. Pneumatosis intestinalis in COVID-19 [J]. BMJ open gastroenterology, 2020, 7(1): 1–5.

[162] XU K, CAI H, SHEN Y, et al. Management of COVID-19: the Zhejiang experience[J]. Journal of Zhejiang University, 2020, 49(2): 147–157.

[163] ABDELHAMID A G, EL-MASRY S S, EL-DOUGDOUG N K. Probiotic *Lactobacillus* and *Bifidobacterium* strains possess safety characteristics, antiviral activities and host adherence factors revealed by genome mining [J]. EPMA journal, 2019, 10(4): 337–350.

[164] NEALON N J, YUAN L, YANG X, et al. Rice bran and probiotics alter the porcine large intestine and serum metabolomes for protection against human rotavirus diarrhea[J]. Frontiers in microbiology, 2017, 8: 1–15.

[165] PARK M S, KWON B, KU S, et al. The efficacy of *Bifidobacterium longum* BORI and *Lactobacillus acidophilus* AD031 probiotic treatment in infants with rotavirus infection[J]. Nutrients, 2017, 9(8): 1–5.

[166] INGLE H, LEE S, AI T, et al. Viral complementation of immunodeficiency confers protection against enteric pathogens via interferon-λ [J]. Nature microbiology, 2019, 4(7): 1120–1128.

[167] REIMER R A. Establishing the role of diet in the microbiota-disease axis [J]. Nature reviews gastroenterology & hepatology, 2019, 16(2): 86–87.

[168] WEN W, QI Z, WANG J. The function and mechanism of *Enterovirus* 71 (EV71)3c protease [J]. Current microbiology, 2020, 77(9): 1968–1975.

[169] WANG J, LI F, WEI H, et al. Respiratory influenza virus infection induces intestinal immune injury via microbiota-mediated Th17 cell-dependent inflammation [J]. Journal of experimental medicine, 2014, 211(12): 2397–2410.

[170] CHEN S Y, TSAI C N, LEE Y S, et al. Intestinal microbiome in children with severe and complicated acute viral gastroenteritis [J]. Scientific reports, 2017, 7 (1): 1–7.

[171] SANDHU K V, SHERWIN E, SCHELLEKENS H, et al. Feeding the microbiota - gut - brain axis: diet, microbiome, and neuropsychiatry [J]. Translational research, 2017, 179: 223–244.

[172] CARRILLO - SALINAS F J, MESTRE L, MECHA M, et al. Gut dysbiosis and neuroimmune responses to brain infection with Theiler's murine encephalomyelitis virus [J]. Scientific reports, 2017, 7(1): 1–16.

第5章
肠道菌群与合成生物学

2022年5月10日，国家发展和改革委员会印发的《“十四五”生物经济发展规划》明确优先发展生物医药、生物农业、生物质替代应用、生物安全四大重点领域，将生物制造列为国家战略性新兴产业。此外，该规划也明确提出从基础研究、创新技术和生物农业产业领域三方面发展合成生物学，将合成生物学视作加快生物经济创新发展的国家战略科技力量之一。

人类肠道微生物组由不同的共生微生物群组成，肠道菌群的基因具有不同关键代谢功能[1]。已发现超过2000种细菌、病毒及真核微生物在肠道中定植[2]。越来越多的证据表明，在心血管、炎症、胃肠道、代谢和神经系统等疾病中，特定的肠道菌群会对宿主健康产生负面影响[3]。肠道中微生物的生态失调可致病，而通过调节肠道菌群能够恢复肠道稳态[4]。许多细菌菌株已被开发为天然益生菌，尤其是乳酸杆菌和双歧杆菌[5]。这些微生物被用于治疗艰难梭菌感染[6]、炎症性疾病[7]，以及焦虑、抑郁和孤独症等神经系统疾病[5]。

合成生物学是一门新的工程学科，在合成生物学中生物体被基因编程从而执行特定功能，合成生物学已被应用于医学、农业和基础生物学等研究领域[8-11]。研究人员使用改编自电气工程的“感知-计算-响应”范式对细胞行为编程[12]，并将生物体设计成表达基因编码的传感器，用于检测细胞内部或外部的特定化学或物理输入。传感器再将化学或物理输入转换为生物输出，并且将生物输出作为工程遗传回路的输入[13, 14]。遗传回路是相互作用的调节分子网络，执行多输入逻辑或存储器等计算。最后，遗传回路通过控制致动基因的活性，对细胞编程以改变其自身或环境状态[11]。

合成生物学为肠道修复提供了一种策略，即通过设计具有治疗功能的细胞实现治疗目标。研究人员可以利用各种遗传工具对传统益生菌针对性修饰，并将其转化为智能药物，克服传统益生菌治疗的缺点，实现个性化治疗[15-17]。这些工程益生菌被预测能够在患病区域定植，且迅速反应并在预设时间内原位产生所需药物，完成治疗功能后自编程死亡。因此，工程益生菌有望参与各种疾病的预防及治疗，例如代谢紊乱[18-22]、肿瘤[23-25]等。在今后的研究中可以通过生物工程师、生物学家、材料科学家、化学家及临床医生的合作，为肠道微生物组相关的慢性疾病提供更好的治疗策略[4]。

本章主要讨论工程益生菌的构建，包括底盘选择、传感器设计、基因回路元件和基因回路设计，以及工程益生菌在治疗肠道细菌疾病、苯丙酮尿症和抑郁症等疾病中的应用与未来展望。

5.1 工程益生菌的构建

在构建益生菌时，必须做出一系列基本设计决策，包括选择生物底盘和肠道疾病生物标志物。为了实现这些目标，研究人员已经开发了多个生物底盘和遗传部件工具包。当目前选择不足时，合成生物学技术还可用于开发新的底盘与遗传部件。

5.1.1 底盘选择

底盘指能够容纳和维持特定功能所需DNA构建体的细胞类型。选择底盘时，一方面需要考虑底盘在胃肠道中的存活能力，确保其能够在肠道内特定位置定植，并达到所需密度；另一方面，还需考虑底盘对DNA转化的适应性、修改底盘基因组工具的可用性，以及传感器、遗传电路与底盘的兼容性等因素[11]。

（1）传统益生菌底盘

因为传统益生菌很容易被设计出来，所以是首选的底盘，并且以往研究已经建立了许多具有不同特性的“现成”底盘。例如，工程化EcN因为其基因组、转录组和代谢特性已被深入研究[26]，被广泛应用于各种生物制药[27, 28]。乳酸菌目前支持各种CRISPR/Cas系统[29, 30]、生物控制系统[31]和生物传感器[32]。双歧杆菌常被用作各种抗癌剂微工厂的原位输送和生产系统[33]，已被开发为益生菌疫苗[34]。大多数革兰氏阴性菌是合成生物学中用于构建新型生物功能体系的主要研究对象，包括*E. coli* CWG308、BW25113[35–38]、Nissle[39, 40]及NGF-1[41]。

（2）新一代益生菌底盘

多形拟杆菌（*Bacteroides thetaiotaomicron*）是近期被开发的益生菌底盘[42]。在拟杆菌属中，多形拟杆菌是第一个被测序的物种[43]。据报道，多形拟杆菌可以减轻肠道炎症，增强机体对病原体入侵的免疫力，并消化重要的膳食营养素，如膳食纤维多糖和宿主聚糖等[44–46]。革兰氏阳性厌氧菌被开发的底盘有干酪乳杆菌[47, 48]、副干酪乳杆菌[49, 50]、植物乳杆菌[32]和乳酸乳球菌[51]，这些益生菌底盘已被普遍认为具有生物安全性，且能够用于疾病治疗[52]。

5.1.2 传感器设计

设计生物传感器简单的方法是通过内源性化学或物理输入响应启动子表达报告基因，而无需对底盘进行额外的工程设计[42, 53]。这种方法虽然简单方便，但由于控制内源性启动子活性的基因调控网络十分复杂并且对其知之甚少，所以会产生假阴性或假阳性的结果。可以通过敲除控制目标启动子的传感器基因进行基因重构[54, 55]，从而破坏控制网络中的关键连接，使传感器性能更加稳定。另一种更可靠的方法是将传感器从不相关的生物体移植到感兴趣的底盘中[56-58]，并优化其遗传编码和表达水平以获得最佳传感器性能[59]。

研究人员还利用天然细菌传感器系统，使其感知肠道出血或炎症标志物并做出反应，以此提高工程微生物的疾病治疗效果[60, 61]。已发现的细菌传感系统有两种，即单组分系统（one component regulatory system，OCS）和双组分系统（two component regulatory system，TCS）[62]。单组分系统是一种主要的细菌信号传导系统，可向细胞传递环境刺激[63]。相反，双组分系统由两种不同的蛋白质组成，一种是感知细胞外刺激的传感器组氨酸激酶（sensor histidine kinase，HK），另一种是将信号从组氨酸激酶转移到下游过程的反应调节因子[64]。

5.1.3 基因回路元件

（1）*启动子*

启动子是指位于结构基因5′端上游的DNA序列，能活化RNA聚合酶，使之与模板DNA准确结合并具有转录起始的特异性。目前因为拟杆菌的转录调控机制与大肠杆菌等特征明确的生物体不同，多数研究集中于对拟杆菌转录调控机制的探索[65-67]。Mimee等对天然拟杆菌启动子进行了工程改造，并在多形拟杆菌中表达异源和同源蛋白，结果发现新构建的启动子极大程度地扩大了组成基因的表达与调节范围[42]。Whitaker等在脆弱拟杆菌（*B. fragilis*）噬菌体基因组中确定了其组成型启动子，并通过在该启动子中引入单个或多个突变构建了一组八个组成型启动子，实现了基因表达与调节范围的扩大[68]。

（2）*核糖体结合位点*

核糖体结合位点（ribosome binding site，RBS）是指

在mRNA的起始AUG密码子上游约8～13个核苷酸处，存在一段由4～9个核苷酸组成的共有序列-AGGAGG-。这一序列能够被16S rRNA通过碱基互补的方式精确识别，从而确保核糖体能够准确地结合到mRNA上启动蛋白质的翻译过程。对多形拟杆菌的RBS文库进行分析显示，富含AT的RBS文库比其他RBS设计更能增强蛋白表达，并且研究表明组成型启动子及RBS的鉴定和表征能够更好地指导研究人员操纵多形拟杆菌的遗传工具包，扩大基因表达水平范围[42, 68]。

（3）转录因子

转录因子（transcription factor，TF）的本质为蛋白质，TF能够结合DNA操作位点激活或抑制启动子转录，是合成生物学中设计逻辑的重要元件。几个基于TF的元件系列已被设计并用于构建一系列遗传逻辑回路。Stanton等设计了一个包含四环素抑制因子（tetracycline repressor，TetR）转录抑制子的正交同源物文库[69]。正交性是指每个TetR同系物仅抑制目标启动子，而不抑制文库中其他TetR同系物的启动子。Stanton等使用此文库构建了一系列“或非”门逻辑电路，并编写了一个新的计算机程序，用于自动遗传逻辑电路设计[69]。

（4）重组酶

重组酶是构建遗传回路的新兴元件，能够产生由重组酶及其在DNA链上的识别位点组成的调节系统。基于识别序列对的特征，重组酶反转或删除某个序列以产生“开启”或“关闭”输出。特别是丝氨酸整合酶在遗传回路设计中具有重要地位，丝氨酸重组酶能够催化DNA之间发生位点特异性重组反应[70]，已被应用于遗传回路中转录控制元件的设计，从而控制RNA聚合酶。这些遗传回路能够形成大量的双输入布尔逻辑门[71, 72]，并可根据其接收输入信号的特性和顺序创建以不同的状态存在的装置（称为状态机器）[73]。

5.1.4 基因回路设计

在传感器和报告基因之间插入遗传逻辑回路，可以提高诊断肠道细菌读数的质量及准确性。工程多形拟杆菌通过生物计算和记忆感知输入进行信号处理，并根据遗传回路做出的“决定”产生输出。输入传感器由组成型和诱导型启动子构成，可检测环境或疾病的生物标志物。在识别出这些生物标志物后，工程多形拟杆菌参与基因调控和遗传记忆的模块可以整合输入，并产生驱动特定和适当的输出功能[74]。

Lundberg等发现膳食硝酸盐会被近端肠道中的微生物群迅速消耗[75]，而炎症性疾病产生的硝酸盐则集中在结肠内[76]。因此，如果同时检测到硝酸盐与大肠生物标志物，那么可能初步诊断为结肠炎。如果同时检测到硝酸盐与小肠生物标志物，则疾病很可能是由饮食引起的而不是炎症。为了执行这种高阶传感功能，构建工程益生菌时需要设计多个传感器和遗传逻辑回路，让其能够同时整合来自这些传感器的生物输入。目前基因回路可以使用诱导表达系统、拨动开关与逻辑门等形式设计。

（1）诱导表达系统

诱导表达系统是遗传回路最简单的形式，如乳糖操纵子、色氨酸操纵子及阿拉伯糖操纵子等，一直被用于构建从基础科学到应用科学的众多生物系统[77]。Saeidi等基于铜绿假单胞菌Ⅰ型群体感应机制设计了一个新型传感系统，该系统包括传感、杀伤和裂解装置，使大肠杆菌能够通过利用合成生物学框架感知和去除致病性铜绿假单胞菌菌株[78]。由组成型的tetR启动子产生与AHL 3OC结合的转录因子LasR12HSL。该转录因子复合物与luxR启动子结合，并激活杀伤及裂解装置，导致大肠杆菌底盘内产生Pyocin S5与裂解E7蛋白。达到阈值浓度后，裂解E7蛋白将大肠杆菌宿主膜穿孔，并释放积累的可溶性蛋白质Pyocin S5。Pyocin S5向目标病原体扩散并破坏其细胞完整性，从而杀死铜绿假单胞菌。这项研究为新型合成生物学抗菌策略提供了基础，该策略可以扩展到幽门螺杆菌及其他病原菌的相关研究中[78]。

（2）拨动开关

拨动开关可以保持设计功能的“开启”状态或“关闭”状态[79]。例如，拨动开关能够被用于检测四硫磺酸（肠道炎症的生物标志物）。在肠道炎症期间，四亚硫酸酯激活β-半乳糖苷酶，使遗传回路从“关闭”状态切换到“开启”状态，以便在诊断中追踪炎症发展[41]。Holowko等将一种基于CRISPRi的诱导遗传回路用于检测霍乱弧菌[80]。该遗传回路能够响应霍乱弧菌特异性CAI-1的表达，在低细胞密度下，CRISPRi抑制报告系统的表达，而在高细胞密度下，单链向导RNA表达受到抑制，通过报告系统进行表达[80]。

（3）逻辑门

为了控制来自多种刺激的多个基因表达，需要设计复杂的遗传回路[81]。为此，研究人员已经构建了生物逻辑门[81]。从简单的“与”门和“或”门到复杂的“或非”门，各种逻辑门都是通过组合许多生物部件构建而成的。其中，“或非”门在硅工业中具有与“与非”门逻辑相当的功能完整性[82]。由于其功能完整性，所有逻辑门都可以通过“或非”门的适当组合实现[82]。RNA分子目前也被用于逻辑门的构建。Arkin等利用工程化质粒pT181的反义RNA介导的转录抑制机制，通过合理的诱变设计了相互正交衰减器，来实现“或非”门的功能[83]。该RNA，还可以创建正交调节RNA，并将其用于构建“或非”门。Chappell等建立了一种新型调控工具：小型转录激活RNAs（small transcription activating RNAs，STARs），STARs能够进行串联组合，并构建出仅基于RNA的转录逻辑门[84]。

5.1.5 定植及其改善策略

益生菌在肠道特定位置定植的能力对于有效诊断或治疗疾病十分必要。例如，对于溃疡性结肠炎患者来说，益生菌在其大肠定植情况会影响具体治疗效果；对于克罗恩病的治疗，益生菌在小肠定植可以改善治疗效果[85]。在结肠中发现数量较多的拟杆菌科（Bacteroidaceae）家族可能更适合治疗溃疡性结肠炎，而在小肠和结肠中发现的乳酸杆菌科成员可能更适合治疗克罗恩病[86]。此外，物种内的特定遗传决定因素可以导致物种不同的定植能力[86]，为构建具有特定定植模式的生物工程底盘奠定了基础。已经开发了各种技术增强工程益生菌的定植能力，这些技术大致可分为三类：表面修饰、遗传修饰和生态位竞争[87]。

（1）表面修饰

细菌的表面改性是指用“保护膜”包裹益生菌，以抵抗胃肠道的恶劣环境，同时促进它们在目标部位的黏附和生长[87]。传统的物理封装技术通常用藻酸盐基质将益生菌微胶囊化，通过避免直接接触使其与恶劣环境隔离开，这种方法显著增加了益生菌对酸性条件和胆汁盐的耐受性。除此之外，也可以使用生物正交介导的细菌递送策略，通过调节益生菌和肠道菌群之间的细菌黏附增强益生菌定植（图5-1）[88]。应用代谢氨基酸工程将叠氮基修饰的D-丙氨酸代谢探针掺入肠道菌群的肽聚糖中，从而实现代谢修饰的肠道细菌与二苯并环辛炔（DBCO）修饰的益生菌原位生物正交偶联[89]。体外和体内研究均表明，即使在复杂的生理环境中，叠氮和DBCO修饰细菌之间生物正交反应的发生也会导致明显的细菌黏附，DBCO修饰的丁酸梭菌（*C. butyricum*）在肠道中也显示出更有效的定植，从而使右旋糖酐硫酸酯钠诱导的结肠炎小鼠的症状显著缓解。

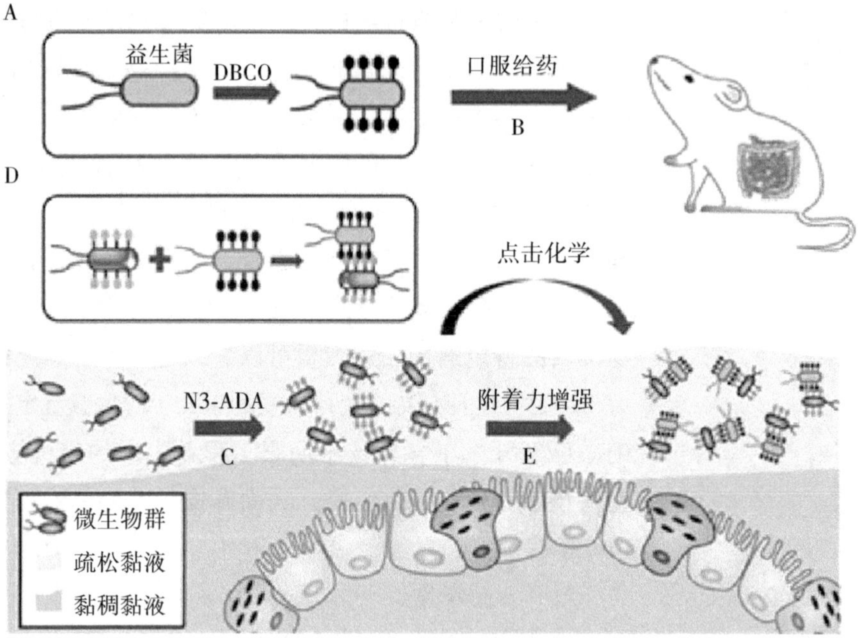

图5-1　通过生物正交介导的细菌递送以增强肠道中益生菌定植的示意图[88]

（2）遗传修饰

最直接的策略是在严格的环境条件下过表达有利于定植的基因，还可以过表达某些参与定植和黏附的细菌蛋白基因，这能够显著提高修饰菌株的黏附能力，并使其表现出与提供基因细菌相似的黏附动力学特征[87]。目前该策略已应用于乳酸乳球菌MG1363中黏液结合蛋白A的过表达。形成功能性的生物膜也是一种常用的策略。作为工程益生菌生物膜的重要组成部分，卷曲纤维是由分泌蛋白CsgA单体的细胞外自组装形成的坚固淀粉样纤维，可以通过功能肽结构域的融合进行工程化和模块化，形成功能化的生物膜以进行黏附。通过使用生物膜集成纳米纤维显示（biofilm - integrated nanofiber display，BIND）共价固定活性三叶因子可将工程细菌的定植能力提高10倍[90]。

(3) 生态位竞争

健康肠道微生物群的稳定性可能源于其多样性和竞争力。因此，可以通过加强提供竞争机会的因素改善定植。特别是对生态位的竞争可以通过饮食调节，即不同碳源对定植有影响。例如，低聚果糖能够作为特定益生菌的碳源促进短链脂肪酸的产生，并且可以到达大肠并选择性地刺激肠道中益生菌的增殖和定植，同时可以塑造微生物群落结构[91]。功能性低聚糖通过促进双歧杆菌和乳酸杆菌在大肠中的定植来发挥其益生元作用，在调节肠道微生物群的同时可防止肠道炎症。

5.2 合成生物学的应用

合成生物学从理解合成生物拟治疗的特定疾病和发展一个药理学假设开始，进而发展出一个精准的药理学假设。这一假设旨在阐明病理生理过程如何在所需作用的特定部位被精准拦截[92]，从而为后续的治疗策略提供坚实的理论基础。目前几类效应器已被设计到细菌中，包括代谢致病毒素的酶途径[93]、生产蛋白质的效应器[94]和生产小分子的效应器[95]。人工合成的生物制剂已被设计应用于皮肤[96]和肿瘤学中的瘤内注射[97]，但最活跃的研究领域仍然是开发用于肠腔[98]的合成生物。

5.2.1 结肠炎

炎症性肠病包括克罗恩病和溃疡性结肠炎，是目前存在的重大公共卫生问题之一，每1000人中就会有1人受到该病影响，但人们对其病因仍知之甚少。炎症性肠病的临床特征主要是大肠或小肠的慢性炎症，表现为腹泻、腹痛、体重减轻和恶心等。在极端情况下，还有可能出现营养不良、脱水、贫血甚至死亡等现象。炎症性肠病被认为起源于相互作用的遗传和环境因素，可能涉及T细胞对共生菌群的异常反应[99]。基于生物学的疗法，如使用肿瘤坏死因子、抗体和重组IL-10治疗[100]，可改善这种疾病。

现如今针对不同疾病已经开发了超过25个基因编码的治疗驱动器。第一类主要是细菌分泌的人类蛋白质，可以增强人类信号网络，主要包括治疗结肠炎的IL-10[101]和抗TNF-α纳米抗体[102]。第二类是增强人类信号网络的细菌蛋白，主要有来自假结核耶尔森氏菌的Toll样受体2的激动剂LcrV蛋白[103]和来自乳酸乳球菌的超氧化物歧化酶[104]，这二者都可减轻结肠炎的症状。另外还有一类是产生治疗性代谢物的遗传途径。

细胞因子IL-10在治疗炎症性肠病的临床试验中显示出较好前景。研究[105]发现通过局部递送基因工程分泌细胞因子的细菌，可以减少IL-10的治疗剂量，口服IL-10分泌型乳酸乳球菌可使葡聚糖硫酸钠处理的小鼠结肠炎减轻50%，并且还能预防小鼠结肠炎。IL-10到达其治疗靶点有两种可能途径[105]：乳球菌可能在管腔中产生IL-10，并且蛋白质可能扩散到上皮或固有层中的反应细胞；或者乳球菌可能被M细胞吸收，主要可能是由于肠淋巴组织中原位产生重组IL-10。这两种途径都可能涉及在炎症中增强的细胞旁转运机制。经过运输后，IL-10可能直接抑制炎症反应，或者通过Tr1细胞等淋巴样细胞产生自分泌物IL-10，以增强上皮细胞的迁移活动，进而增强组织的修复能力。

综上，由食品级细菌原位合成的具有成本效益的局部递送治疗剂，为炎症性肠病治疗提供了潜在的临床应用前景，尤其是可将其用于全身治疗。

5.2.2 苯丙酮尿症

苯丙酮尿症（phenylketonuria，PKU）是一种先天性代谢错误疾病，由编码苯丙氨酸（phenylalanine，Phe）羟化酶（一种负责将Phe转化为酪氨酸的酶）的基因遗传缺陷引起。如果不尽早发现并通过严格的Phe控制饮食治疗，PKU可导致不可逆转的神经损伤[106]。Phe是大多数膳食蛋白质的组成成分，可以从人体循环中循环回肠道，因此可以通过口服可降解Phe的肠道限制性合成生物学制剂治疗PKU[107]。

为了创造一种适合治疗PKU的合成生物，Synlogic公司的菌株工程队构建了Phe降解菌大肠杆菌Nissle 1917的衍生物SYNB1618，将两种Phe降解途径整合进大肠杆菌Nissle（图5-2）。第一种途径使用苯丙氨酸解氨酶（phenylalanine ammonia lyase，PAL）将Phe转化为反式肉桂酸（trans-cinnamic acid，TCA），TCA是一种无害的代谢物。PAL是一种胞质蛋白，需要将Phe转运到细胞中，因此提高了高亲和力Phe转运蛋白（Phenylalanine transportprotein，PheP）的表达，以增加Phe的细胞内可用性。第二种Phe降解途径使用L-氨基酸氧化酶（L-amino-acid oxidase，LAAO），这是一种可将Phe转化为苯丙酮酸的膜相关酶。这些编码酶和转运蛋白的基因，在诱导启动子的精确调控下，被巧妙地整合至大肠杆菌Nissle的染色体中。这一策略旨在实现在高密度、大规模生长阶段对人类所需剂量的精确控制，同时确保遗传完整性的维持，为后续的应用奠定了坚实基础。最后，SYNB1618缺失了编码4-羟基四氢吡啶合成酶的dapA基因，使工程细菌依赖于外源性二氨基庚二酸进行细胞壁生物合成和生长来实现生物防护[18]。

研究发现，SYNB1618具有良好的安全性和耐受性。在较高剂量下使用SYNB1618，可观察到轻度至中度的胃肠道不良事件[92]。与临床数据一致，血浆和尿液中菌株特异性Phe代谢物的剂量也随之增加，这为菌株在胃肠道消耗Phe的机制提供了证明。

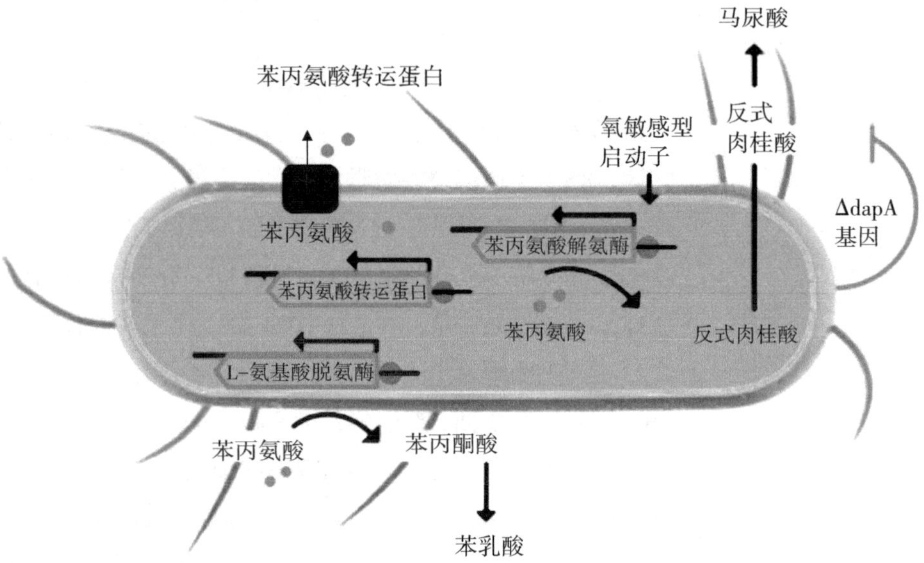

图5-2 大肠杆菌Nissle的衍生物SYNB1618代谢苯丙氨酸示意图[92]

5.2.3 抑郁症

抑郁症的多面性和异质性给精准治疗带来了许多挑战。多种微生物组产生的代谢物通过肠道菌群和神经功能之间的肠-脑轴与宿主信号和代谢途径相互作用。因此，肠道代谢物的生物传感器检测提供了量化微生物对抑郁症贡献的潜力[108]。工程生物传感器可用于监测可能影响抑郁症的肠道代谢变化（图5-3），如可用于监测一氧化氮（NO）和四亚硫酸（$S_4O_6^{2-}$）等炎性生物标志物、色氨酸和短链脂肪酸等的代谢产物，以及pH和O_2等信号分子。

NO作为一种重要的炎症生物标志物[109]，其水平的升高常常与上皮细胞的炎症反应密切相关。在这种炎症环境下，诱导性一氧化氮合酶（inducible nitric oxide synthase，iNOS）的表达会受到显著刺激。鉴于这一紧密的联系，iNOS有可能成为抑郁症的一个信息性生物标志物。在细菌对NO的应激反应中，存在两种关键的调节因子，即一氧化氮敏感调节因子（nitric oxide sensing regulator，NorR）和一氧化氮敏感转录抑制因子（nitric oxide-sensitive transcriptional repressor，NsrR），它们共同发挥着至关重要的作用。将NorR与DNA重组酶的调节偶联，利用了天然NO传感，从而永久记录与小鼠回肠的外植体培养中检测肠道炎症相关的NO信号[110]。为了优化NO生物传感器，引入了正反馈回路微调大肠杆菌Nissle 1917中的NorR水平[111]，通过正反馈调节，NO也诱导NorR表达，促进活化NorR的增加，这为生物传感器提供了更大的动态范围。

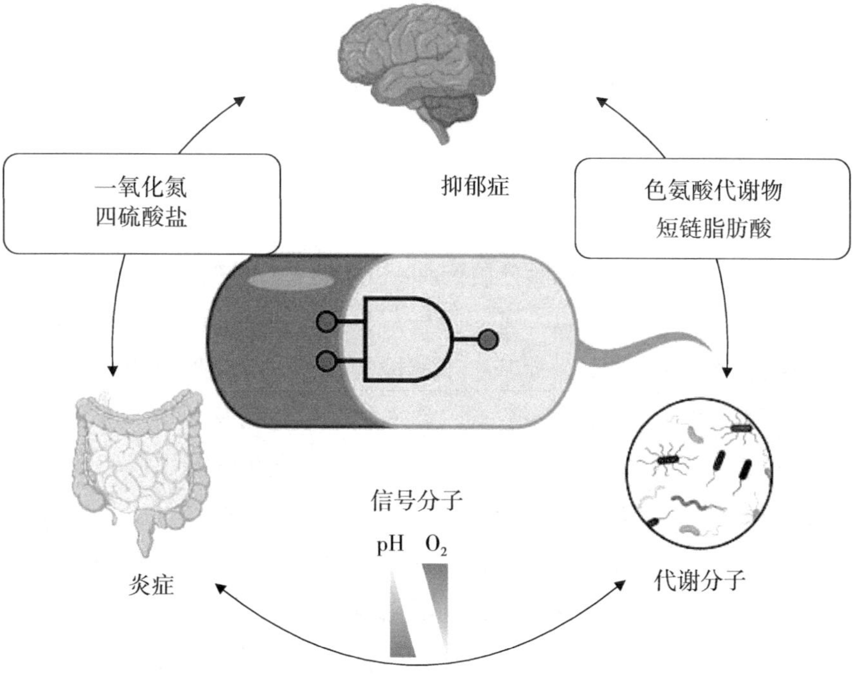

图5-3 通过肠-脑轴用生物传感器监测抑郁症示意图[108]

色氨酸代谢物作为抑郁症的潜在生物标志物，色氨酸阻遏蛋白的蛋白质工程可以快速检测色氨酸衍生代谢物（如吲哚）的产生。吲哚-3-醛（indole-3-aldehyde，I3A）可以激活芳香烃受体（aryl hydrocarbon receptor，

AhR）并调节中枢神经系统炎症[112]。为了检测I3A，基于细菌双组分信号系统的信号部件设计了一种生物传感器。工程化组氨酸激酶是一种杂交蛋白，包含有I3A可结合的Per-Arnt-Sim结构域[113]。在启动子区域内的优化改善了动态范围，并在检测I3A时使信号发生了约35倍的变化。另外，pH和O_2与抑郁症生物标志物传感器相结合，具有复杂电路的生物传感器可以识别产生抑郁症相关生物标志物的特定微生物生态位。

肠-脑轴的最新研究表明，肠道和大脑中的代谢物水平之间存在定性相关性。随着技术的进一步发展，肠道代谢物的生物传感器监测可能会形成针对心理健康相关疾病的非侵入性、精确且个性化的药物[114]。但是，生物传感器的应用也会受到代谢物信号的特异性、信号的动态范围、启动子泄漏以及与其他途径形成潜在串扰的挑战。

5.3 展望

5.3.1 功能性治疗回路和生物防护回路的优化

大肠杆菌Nissle和乳酸菌继续主导着当前的工程益生菌底盘。基于拟杆菌和双歧杆菌的活细菌疗法开始崭露头角，而基于丁酸梭菌和布拉氏酵母菌的活细菌疗法还处于起步阶段。新一代益生菌底盘必须丰富合成生物学组件包，除了需要更简单、通用的基因操作技术来降低技术应用门槛和成本外，还需要扩展可用的底盘并仔细评估每个底盘的安全性和机动性。未来还可以利用机箱的特性对症设计活细菌疗法[115]。例如考虑使用更能适应小肠恶劣定植环境的乳酸菌进行克罗恩病的治疗；考虑使用大肠杆菌Nissle或双歧杆菌发挥其肿瘤靶向作用，以此治疗结肠肿瘤。由于定制这些益生菌的最终目的是服务于患者，所以益生菌商业化的可行性是未来必须考虑的因素之一，例如低成本大规模生产的可行性、设计时间的可定制性、材料成本的可控制性、各种工业过程的耐受性、给药封装形式和保质期等的可行性。

合成生物学的功能基因回路设计可以使工程益生菌灵敏地检测外部信号的变化并产生预设效果的功能，同时也赋予工程益生菌记忆[116]、逻辑运算[117]和围捕[118]等功能。所以有必要利用更多的正交元素提供系统的可扩展性，丰富遗传工具集，例如TetR家族的模块化工程[119]可以对更多的信号执行复杂的计算[120]。因此，工程益生菌可以处理连续的环境信号，并且有报道显示其可以监测肠道中的炎症反应分子[117]。这种策略可以使工程益生菌更加智能，使它们能够自动确定肠道疾病的进展状态并采取适当的治疗策略。

另外，多种生物防护策略的联合应用大大降低了逃逸率，使生物防护系统变得极其稳定。多层保护、多信号响应和多致死基因输出的策略有望提供更高的稳定性。然而，在设计中也应考虑代谢负担，特别是注意工程益生菌的主要任务是发出反应信号，而不是专注于自杀。因此，除了在基因组上编码系统外，生态位竞争可能为减轻负担提供了很好的选择[121]。所以需要新的生物防逃逸策略克服现有的缺陷。例如，对于诱导自杀回路的诱导剂必须是对身体或肠道菌群无害的非抗生素化学物质，并且必须仔细评估这些化学物质的药代动力学。

5.3.2 遗传稳定性和功能稳定性的维持

无论是功能性治疗回路还是生物防护回路，这些工程遗传元件的运作都不可避免地依赖于宿主的代谢资源，所以会造成不同程度的代谢负担，进而引起不同程度的毒性。宿主自然进化为失去外来基因或变异外源线路以获得适应性优势，这在无抗生素肠道中更为常见，最终导致突变体占据整个种群。因此，在设计中应保证工程电路的遗传稳定性和功能稳定性。基于此，可以采用将功能元件整合到基因组[122]、内源性质粒中[123]或使用毒素-抗毒素系统[124]和细菌素[125]稳定质粒遗传的方法。但是这种方法需要平衡由此产生的电路性能牺牲[126]以及固有基因表达的影响[127]。所以可以选择改良和培育低突变率的宿主[128]，人为减慢进化速度[129]以增强功能稳定性，或者采用遗传冗余策略来防止单拷贝基因功能丧失[18]。同时，采取必要措施，如靶向回收胞内蛋白资源[130]及建立模块化压力反馈传感器[131]，从而减少功能回路的代谢负担和毒性，避免突变并为昂贵的基因提供有针对性的保护。

5.3.3 人工智能和合成生物学与食品科学结合改善肠道菌群

由于人工智能（artificial intelligence，AI）技术的不断进步，通过元数据的管理，人工智能使人们能够更好地了解健康肠道微生物群和失调患者的微生物群。伴随着合成生物学的出现，这些健康的肠道遗传信息可用于制造工程菌株，将其用作膳食补充剂[132]服用时可以恢复人体肠道的正常微生物群落[1]。工程菌株作为有效药物，已被证明可以产生营养物质、酶、抗氧化剂、化学分子等，它们可以作为益生菌添加到人类饮食中[133]。

目前，现有的合成生物学工具尚局限于一次性改造一种微生物。因此，当前市场上的益生菌大多以富含特定细菌种类为主，这反映了当前技术的实际应用水平。然而，只有考虑到这些微生物与宿主之间的动态相互作用的情况下，合成生物学才能产生多种有益微生物的合成形式，才能潜在地影响人类健康。此外，用于生产工程菌的合成生物学方法的成本和可扩展性问题需要解决。这些膳食补充剂结合适当的食物类型，可以容纳有益的微生物，并促进它们作为智能生物反应器的生长和发育。虽然人工智能和合成生物学已经达到了通过肠道改善人类健康的重要地位，但这些技术仍在不断发展，只有经过有关食品监管当局的彻底审查和批准后才能被采用，在诊断和治疗方面的安全性确定后，这些工具才能作为医疗行业的主要工具。因此，未来人工智能与食品科学的结合将有助于维持健康的肠道菌群，从而增进人类健康和福祉。

5.3.4 多细胞群体黏附

多细胞群体黏附可能为未来的活体疗法提供一种新的方法[134]。用特定的合成黏附素纳米抗体修饰的大肠杆菌Nissle可用于控制预定结构中不同大肠杆菌Nissle衍生物的结合。例如，使用一个在其表面展示多个选择抗原的诱饵菌株，目标菌株与表面展示的同源抗原的结合将赋予群体黏附性[135]。该策略实现了单一物种的群体黏附，未来可能产生多物种细胞群落。此方法代表了一种灵活的工程策略，也就是说，从群体的角度看，我们可以使不同的工程益生菌协同工作，而不需要每种益生菌都能达到完美的性能。例如，具有最强大定植能力的益生菌——拟杆菌，可以专门为其他具有治疗能力，但定植能力较差的益生菌提供着陆点。未来，该方向的研究成果有望超越目前的单一工程底盘治疗策略，并在工程化群落水平调节肠道群落属性。

合成生物治疗作为一种新的治疗方式，具有巨大的发展前景。合成生物制剂是可以根据生理或病理信号在人体内的作用部位进行效价设计并对局部条件做出反应的活体机器。合成生物制剂的出现实现了有针对性地提供治疗，并在正确的时间、正确的地点提供正确的效应器，从而改善医疗效果。在未来几十年，合成生物制剂将对疾病诊断、疫苗生产和治疗方法革新产生巨大影响。

（范兢文　姜玉超　韩华雯*）

参考文献

[1] KUMAR P, SINHA R, SHUKLA P. Artificial intelligence and synthetic biology approaches for human gut microbiome[J]. Critical reviews in food science and nutrition, 2022, 62(8): 2103-2121.

[2]THURSBY E,JUGE N. Introduction to the human gut microbiota[J]. Biochemical journal,2017,474(11):1823–1836.

[3]SHREINER A B,KAO J Y,YOUNG V B. The gut microbiome in health and in disease[J]. Current opinion in gastroenterology,2015,31(1):69–75.

[4]TAN X,YANG Y L,LI X,et al. Multi-metabolism regulation insights into nutrients removal performance with adding heterotrophic nitrification-aerobic denitrification bacteria in tidal flow constructed wetlands[J]. Science of the total environment,2021,796:1–12.

[5]WANG H,LEE I S,BRAUN C,et al. Effect of probiotics on central nervous system functions in animals and humans:a systematic review[J].Journal of neurogastroenterol motil,2016,22(4):589–605.

[6]CHOI H H,CHO Y S. Fecal microbiota transplantation:current applications,effectiveness,and future perspectives[J]. Clinical endocrinology,2016,49(3):257–265.

[7]BRON P A,KLEEREBEZEM M,BRUMMER R J,et al. Can probiotics modulate human disease by impacting intestinal barrier function?[J]. British journal of nutrition,2017,117(1):93–107.

[8]GORDLEY R M,BUGAJ L J,LIM W A. Modular engineering of cellular signaling proteins and networks[J]. Current opinion in structural biology,2016,39:106–114.

[9]BROPHY J A,VOIGT C A. Principles of genetic circuit design[J]. Nature methods,2014,11(5):508–520.

[10]SMANSKI M J,ZHOU H,CLAESEN J,et al. Synthetic biology to access and expand nature's chemical diversity[J]. Nature reviews microbiology,2016,14(3):135–149.

[11]LANDRY BRIAN P,TABOR JEFFREY J. Engineering diagnostic and therapeutic gut bacteria[J]. Microbiology spectrum,2017,5:1–22.

[12]TABOR J J,GROBAN E S,VOIGT C A. Systems biology and biotechnology of *Escherichia coli*[M]. New York:Springer,2009.

[13]OLSON E J,HARTSOUGH L A,LANDRY B P,et al. Characterizing bacterial gene circuit dynamics with optically programmed gene expression signals[J]. Nature

methods,2014,11(4):449–455.

[14]CASTILLO-HAIR S M,IGOSHIN O A,TABOR J J. How to train your microbe:methods for dynamically characterizing gene networks[J]. Current opinion in microbiology,2015,24:113–123.

[15]SINGH T P,NATRAJ B H.Next-generation probiotics:a promising approach towards designing personalized medicine[J]. Critical reviews in microbiology,2021,47(4):479–498.

[16]PEDROLLI D B,RIBEIRO N V,SQUIZATO P N,et al. Engineering microbial living therapeutics:the synthetic biology toolbox[J]. Trends in biotechnology,2019,37(1):100–115.

[17]DOSOKY N S,MAY-ZHANG L S,DAVIES S S. Engineering the gut microbiota to treat chronic diseases[J]. Applied microbiology and biotechnology,2020,104:7657–7671.

[18]KERRY R G,DAS G,GOLLA U,et al. Engineered probiotic and prebiotic nutraceutical supplementations in combating non-communicable disorders: a review[J]. Current pharmaceutical biotechnology,2022,23(1):72–97.

[19]AGGARWAL N,BREEDON A M,DAVIS C M,et al. Engineering probiotics for therapeutic applications:recent examples and translational outlook[J]. Current opinion in biotechnology,2020,65:171–179.

[20]ZHAO R,LI Z,SUN Y,et al. Engineered *Escherichia coli* Nissle 1917 with urate oxidase and an oxygen-recycling system for hyperuricemia treatment[J]. Gut microbes,2022,14(1):1–20.

[21]ADOLFSEN K J,CALLIHAN I,MONAHAN C E,et al. Improvement of a synthetic live bacterial therapeutic for phenylketonuria with biosensor-enabled enzyme engineering[J]. Nature communications,2021,12(1):1–13.

[22]PUURUNEN M K,VOCKLEY J,SEARLE S L,et al. Safety and pharmacodynamics of an engineered *E. coli* Nissle for the treatment of phenylketonuria:a first-in-human phase 1/2a study[J]. Nature metabolism,2021,3(8):1125–1132.

[23]CHOWDHURY S,CASTRO S,COKER C,et al. Programmable bacteria induce durable tumor regression and systemic antitumor immunity[J]. Nature medicine,2019,25(7):1057–1063.

[24]GURBATRI C R,RADFORD G A,VRBANAC L,et al. Engineering tumor-colonizing *E. coli* Nissle 1917 for detection and treatment of colorectal neoplasia[J]. Nature communications,2024,15(1):1-13.

[25]ABEDI M H,YAO M S,MITTELSTEIN D R,et al. Ultrasound-controllable engineered bacteria for cancer immunotherapy[J]. Nature communications,2022,13(1):1-11.

[26]FARASAT I,KUSHWAHA M,COLLENS J,et al. Efficient search,mapping,and optimization of multi-protein genetic systems in diverse bacteria[J]. Molecular systems biology,2014,10(6):1-18.

[27]ISABELLA V M,HA B N,CASTILLO M J,et al. Development of a synthetic live bacterial therapeutic for the human metabolic disease phenylketonuria[J]. Nature biotechnology,2018,36(9):857-864.

[28]KURTZ C B,MILLET Y A,PUURUNEN M K,et al. An engineered *E. coli* Nissle improves hyperammonemia and survival in mice and shows dose-dependent exposure in healthy humans[J]. Science translational medicine,2019,11(475):1-14.

[29]BOSMA E F,FORSTER J,NIELSEN A T. Lactobacilli and pediococci as versatile cell factories-Evaluation of strain properties and genetic tools[J]. Biotechnology advances,2017,35(4):419-442.

[30]MIERAU I,KLEEREBEZEM M. 10 years of the nisin-controlled gene expression system(NICE)in *Lactococcus lactis*[J]. Applied microbiology and biotechnology,2005,68(6):705-717.

[31]SHETTY R P,ENDY D,KNIGHT T F. Engineering biobrick vectors from biobrick parts[J]. Journal of biological engineering,2008,2:1-12.

[32]HAN W,MERCENIER A,AIT-BELGNAOUI A,et al. Improvement of an experimental colitis in rats by lactic acid bacteria producing superoxide dismutase[J]. Inflammatory bowel diseases,2006,12(11):1044-1052.

[33]CLAESEN J,FISCHBACH M A. Synthetic microbes as drug delivery systems[J]. ACS synthetic biology,2015,4(4):358-364.

[34]KARLSKÅS I L,MAUDAL K,AXELSSON L,et al. Heterologous protein secretion in lactobacilli with modified pSIP vectors[J]. PLoS one,2014,9(3):1-9.

[35]PATON A W,JENNINGS M P,MORONA R,et al. Recombinant probiotics

for treatment and prevention of enterotoxigenic *Escherichia coli* diarrhea[J]. Gastroenterology,2005,128(5):1219–1228.

[36]FOCARETA A,PATON J C,MORONA R,et al. A recombinant probiotic for treatment and prevention of cholera[J]. Gastroenterology,2006,130(6):1688–1695.

[37]PICKARD J M,MAURICE C F,KINNEBREW M A,et al. Rapid fucosylation of intestinal epithelium sustains host - commensal symbiosis in sickness [J]. Nature, 2014,514(7524):638–641.

[38]CASTAGLIUOLO I,BEGGIAO E,BRUN P,et al. Engineered *E. coli* delivers therapeutic genes to the colonic mucosa[J]. Gene therapy,2005,12(13):1070–1078.

[39]CHEN H L,LAI Y W,CHEN C S,et al. Probiotic *Lactobacillus casei* expressing human lactoferrin elevates antibacterial activity in the gastrointestinal tract[J]. Biometals,2010,23:543–554.

[40]DUAN F,MARCH J C. Engineered bacterial communication prevents *Vibrio cholerae* virulence in an infant mouse model[J]. Proceedings of the national academy of sciences,2010,107(25):11260–11264.

[41]RIGLAR D T,GIESSEN T W,BAYM M,et al. Engineered bacteria can function in the mammalian gut long-term as live diagnostics of inflammation[J]. Nature biotechnology,2017,35(7):653–658.

[42]MIMEE M,TUCKER A C,VOIGT C A,et al. Programming a human commensal bacterium,*Bacteroides thetaiotaomicron*,to sense and respond to stimuli in the murine gut microbiota[J]. Cell systems,2015,1(1):62–71.

[43]XU J,BJURSELL M K,HIMROD J,et al. A genomic view of the human-*Bacteroides thetaiotaomicron* symbiosis[J]. Science,2003,299(5615):2074–2076.

[44]KELLY D,CAMPBELL J I,KING T P,et al. Commensal anaerobic gut bacteria attenuate inflammation by regulating nuclear - cytoplasmic shuttling of PPAR - gamma and RelA[J]. Nature immunology,2004,5(1):104–112.

[45]WRZOSEK L,MIQUEL S,NOORDINE M L,et al. *Bacteroides thetaiotaomicron* and *Faecalibacterium prausnitzii* influence the production of mucus glycans and the development of goblet cells in the colonic epithelium of a gnotobiotic model rodent [J]. BMC biology,2013,11:1–13.

[46]PORTER N T,LUIS A S,MARTENS E C. *Bacteroides thetaiotaomicron*[J].

Trends in microbiology, 2018, 26(11): 966–967.

[47] CARROLL I M, ANDRUS J M, BRUNO-BÁRCENA J M, et al. Anti-inflammatory properties of *Lactobacillus gasseri* expressing manganese superoxide dismutase using the interleukin 10-deficient mouse model of colitis[J]. American journal of physiology-gastrointestinal and liver physiology, 2007, 293(4): 729–738.

[48] DUAN F F, LIU J H, MARCH J C. Engineered commensal bacteria reprogram intestinal cells into glucose-responsive insulin-secreting cells for the treatment of diabetes[J]. Diabetes, 2015, 64(5): 1794–1803.

[49] ROSBERG-CODY E, STANTON C, O'MAHONY L, et al. Recombinant lactobacilli expressing linoleic acid isomerase can modulate the fatty acid composition of host adipose tissue in mice[J]. Microbiology, 2011, 157(2): 609–615.

[50] KOO O K, AMALARADJOU M A, BHUNIA A K. Recombinant probiotic expressing Listeria adhesion protein attenuates *Listeria monocytogenes* virulence in *vitro* [J]. PLoS one, 2012, 7(1): 1–14.

[51] HERNANDEZ-VALDES J A, SOLOPOVA A, KUIPERS O P. Development of *Lactococcus lactis* biosensors for detection of diacetyl[J]. Frontiers in microbiology, 2020, 11: 1–15.

[52] PONTES D S, DE AZEVEDO M S, CHATEL J M, et al. *Lactococcus lactis* as a live vector: heterologous protein production and DNA delivery systems[J]. Protein expression and purification, 2011, 79(2): 165–175.

[53] DROUAULT S, ANBA J, CORTHIER G. *Streptococcus thermophilus* is able to produce a beta-galactosidase active during its transit in the digestive tract of germ-free mice[J]. Applied and environmental microbiology, 2002, 68(2): 938–941.

[54] CHAN L Y, KOSURI S, ENDY D. Refactoring bacteriophage T7[J]. Molecular systems biology, 2005, 1(1): 1–10.

[55] TEMME K, ZHAO D, VOIGT C A. Refactoring the nitrogen fixation gene cluster from *Klebsiella oxytoca* [J]. Proceedings of the national academy of sciences, 2012, 109(18): 7085–7090.

[56] DAEFFLER K N, GALLEY J D, SHETH R U, et al. Engineering bacterial thiosulfate and tetrathionate sensors for detecting gut inflammation[J]. Molecular systems biology, 2017, 13(4): 1–13.

[57]TABOR J J, LEVSKAYA A, VOIGT C A. Multichromatic control of gene expression in *Escherichia coli*[J]. Journal of molecular biology, 2011, 405(2): 315–324.

[58]RAMAKRISHNAN P, TABOR J J. Repurposing *Synechocystis* PCC6803 uirs-uirr as a uv-violet/green photoreversible transcriptional regulatory tool in *E. coli*[J]. ACS synthetic biology, 2016, 5(7): 733–740.

[59]SCHMIDL S R, SHETH R U, WU A, et al. Refactoring and optimization of light-switchable *Escherichia coli* two-component systems[J]. ACS synthetic biology, 2014, 3(11): 820–831.

[60]SCHMIDT F, ZIMMERMANN J, TANNA T, et al. Noninvasive assessment of gut function using transcriptional recording sentinel cells[J]. Science, 2022, 376(6594): 1–17.

[61]INDA M E, BROSET E, LU T K, et al. Emerging frontiers in microbiome engineering[J]. Trends in immunology, 2019, 40(10): 952–973.

[62]RIGLAR D T, SILVER P A. Engineering bacteria for diagnostic and therapeutic applications[J]. Nature reviews microbiology, 2018, 16(4): 214–225.

[63]ULRICH L E, KOONIN E V, ZHULIN I B. One-component systems dominate signal transduction in prokaryotes[J]. Trends microbiol, 2005, 13(2): 52–56.

[64]ZSCHIEDRICH C P, KEIDEL V, SZURMANT H. Molecular mechanisms of two-component signal transduction[J]. Journal of molecular biology, 2016, 428(19): 3752–3775.

[65]BAYLEY D P, ROCHA E R, SMITH C J. Analysis of *cepA* and other *Bacteroides fragilis* genes reveals a unique promoter structure[J]. Fems microbiology letters, 2000, 193(1): 149–154.

[66]VINGADASSALOM D, KOLB A, MAYER C, et al. An unusual primary sigma factor in the Bacteroidetes phylum[J]. Molecular microbiology, 2005, 56(4): 888–902.

[67]MASTROPAOLO M D, THORSON M L, STEVENS A M. Comparison of *bacteroides thetaiotaomicron* and *Escherichia coli* 16S rRNA gene expression signals[J]. Microbiology, 2009, 155(8): 2683–2693.

[68]WHITAKER W R, SHEPHERD E S, SONNENBURG J L. Tunable expression tools enable single-cell strain distinction in the gut microbiome[J]. Cell, 2017, 169

(3):538–546.

[69]STANTON B C, NIELSEN A A, TAMSIR A, et al. Genomic mining of prokaryotic repressors for orthogonal logic gates[J]. Nature chemical biology, 2014, 10(2):99–105.

[70]ARGOS P, LANDY A, ABREMSKI K, et al. The integrase family of site-specific recombinases: regional similarities and global diversity[J]. EMBO journal, 1986, 5(2): 433–440.

[71]SIUTI P, YAZBEK J, LU T K. Synthetic circuits integrating logic and memory in living cells[J]. Nature biotechnology, 2013, 31(5):448–452.

[72]BONNET J, YIN P, ORTIZ M E, et al. Amplifying genetic logic gates[J]. Science, 2013, 340(6132):599–603.

[73]ROQUET N, SOLEIMANY A P, FERRIS A C, et al. Synthetic recombinase-based state machines in living cells[J]. Science, 2016, 353(6297):1–14.

[74]LAI Y, HAYASHI N, LU T K. Engineering the human gut commensal *Bacteroides thetaiotaomicron* with synthetic biology[J]. Current opinion in chemical biology, 2022, 70:1–10.

[75]LUNDBERG J O, GOVONI M. Inorganic nitrate is a possible source for systemic generation of nitric oxide[J]. Free radical biology and medicine, 2004, 37(3): 395–400.

[76]WINTER S E, WINTER M G, XAVIER M N, et al. Host-derived nitrate boosts growth of *E. coli* in the inflamed gut[J]. Science, 2013, 339(6120):708–711.

[77]STETZ M A, CARTER M V, WAND A J. Optimized expression and purification of biophysical quantities of Lac repressor and Lac repressor regulatory domain[J]. Protein expression and purification, 2016, 123:75–82.

[78]SAEIDI N, WONG C K, LO T M, et al. Engineering microbes to sense and eradicate *Pseudomonas aeruginosa*, a human pathogen[J]. Molecular systems biology, 2011, 7(1):1–11.

[79]GARDNER T S, CANTOR C R, COLLINS J J. Construction of a genetic toggle switch in *Escherichia coli*[J]. Nature, 2000, 403(6767):339–342.

[80]HOLOWKO M B, WANG H, JAYARAMAN P, et al. Biosensing *vibrio cholerae* with genetically engineered *Escherichia coli*[J]. ACS synthetic biology, 2016, 5(11):

1275–1283.

[81]SAYUT D J, NIU Y, SUN L. Engineering the logical properties of a genetic AND gate[J]. Methods in molecular biology, 2011, 743: 175–184.

[82]GOÑI-MORENO A, AMOS M. A reconfigurable NAND/NOR genetic logic gate[J]. BMC systems biology, 2012, 6: 1–11.

[83]LUCKS J B, QI L, MUTALIK V K, et al. Versatile RNA-sensing transcriptional regulators for engineering genetic networks[J]. Proceedings of the national academy of sciences, 2011, 108(21): 8617–8622.

[84]CHAPPELL J, TAKAHASHI M K, LUCKS J B. Creating small transcription activating RNAs[J]. Nature chemical biology, 2015, 11(3): 214–220.

[85]CONRAD K, ROGGENBUCK D, LAASS M W. Diagnosis and classification of ulcerative colitis[J]. Autoimmunity reviews, 2014, 13(4/5): 463–466.

[86]DONALDSON G P, LEE S M, MAZMANIAN S K. Gut biogeography of the bacterial microbiota[J]. Nature reviews microbiology, 2016, 14(1): 20–32.

[87]HUANG Y, LIN X, YU S, et al. Intestinal Engineered probiotics as living therapeutics: chassis selection, colonization enhancement, gene circuit design, and biocontainment[J]. ACS synthetic biology, 2022, 11(10): 3134–3153.

[88]SONG W F, YAO W Q, CHEN Q W, et al. In situ bioorthogonal conjugation of delivered bacteria with gut inhabitants for enhancing probiotics colonization[J]. ACS central science, 2022, 8(9): 1306–1317.

[89]LIN L, DU Y, SONG J, et al. Imaging commensal microbiota and pathogenic bacteria in the gut[J]. Accounts of chemical research, 2021, 54(9): 2076–2087.

[90]NGUYEN P Q, BOTYANSZKI Z, TAY P K R, et al. Programmable biofilm-based materials from engineered curli nanofibres[J]. Nature communications, 2014, 5(1): 1–10.

[91]ZHANG N, JIN M, WANG K, et al. Functional oligosaccharide fermentation in the gut: improving intestinal health and its determinant factors-a review[J]. Carbohydrate polymers, 2022, 284: 1–13.

[92] BRENNAN A M. Development of synthetic biotics as treatment for human diseases[J]. Synthetic biology, 2022, 7(1): 1–7.

[93] ADOLFSEN K J, CALLIHAN I, MONAHAN C E, et al. Improvement of a

synthetic live bacterial therapeutic for phenylketonuria with biosensor-enabled enzyme engineering[J]. Nature communications, 2021, 12(1): 6215–6228.

[94] BRAAT H, ROTTIERS P, HOMMES D W, et al. A phase I trial with transgenic bacteria expressing interleukin-10 in crohn's disease[J]. Clinical gastroenterology and hepatology, 2006, 4(6): 754–759.

[95] LEVENTHAL D S, SOKOLOVSKA A, LI N, et al. Immunotherapy with engineered bacteria by targeting the STING pathway for anti-tumor immunity[J]. Nature communications, 2020, 11: 1–15.

[96] DODDS D, BOSE J L, DENG M D, et al. Controlling the growth of the skin commensal *Staphylococcus epidermidis* using D-alanine auxotrophy[J]. Msphere, 2020, 5(3): 360–380.

[97] WEST K A, FISHER A, LEVENTHAL D, et al. Metabolic modulation of the tumor microenvironment using Synthetic Biotic™ Medicines[J]. Cancer research, 2018, 78(13): 2920–2031.

[98] LIMAYE S A, SONIS S T, CILLI F, et al. Phase 1b, multicenter, single-blinded, placebo-controlled, sequential dose-escalation study to assess the safety and tolerability of topically applied AG013 in subjects with locally advanced head and neck cancer receiving induction chemotherapy[J]. Journal of clinical oncology, 2013, 119(24): 4268–4276.

[99] RATH H C, HERFARTH H H, IKEDA J S, et al. Normal luminal bacteria, especially *Bacteroides* species, mediate chronic colitis, gastritis, and arthritis in HLA-B27/human beta2 microglobulin transgenic rats[J]. Journal of clinical investigation, 1996, 98(4): 945–953.

[100] VANDEVENTER S J H, ELSON C O, Fedorak R N. Multiple doses of intravenous interleukin 10 in steroid-refractory Crohn's disease[J]. Gastroenterology, 1997, 113(2): 383–389.

[101] STEIDLER L, HANS W, SCHOTTE L, et al. Treatment of murine colitis by *Lactococcus lactis* secreting interleukin-10[J]. Science, 2000, 289(5483): 1352–1355.

[102] VANDENBROUCKE K, DE HAARD H, BEIRNAERT E, et al. Orally administered *L. lactis* secreting an anti-TNF nanobody demonstrate efficacy in chronic colitis[J]. Mucosal immunology, 2010, 3(1): 49–56.

[103] FOLIGNE B, DESSEIN R, MARCEAU M, et al. Prevention and treatment of colitis with *Lactococcus lactis* secreting the immunomodulatory *Yersinia* LcrV protein [J]. Gastroenterology, 2007, 133(3): 862–874.

[104] HAN W, MERCENIER A, AIT-BELGNAOUI A, et al. Improvement of an experimental colitis in rats by lactic acid bacteria producing superoxide dismutase [J]. Inflammatory bowel diseases, 2006, 12(11): 1044–1052.

[105] GROUX H, OGARRA A, BIGLER M, et al. A CD_4^+ T-cell subset inhibits antigen-specific T-cell responses and prevents colitis [J]. Nature, 1997, 389(6652): 737–742.

[106] VOCKLEY J, ANDERSSON H C, ANTSHEL K M, et al. Phenylalanine hydroxylase deficiency: diagnosis and management guideline [J]. Genetics in medicine, 2014, 16(2): 188–200.

[107] CHANG T M S, BOURGET L, LISTER C. A new theory of enterorecirculation of amino acids and its use for depleting unwanted amino acids using oral enzyme-artificial cells, as in removing phenylalanine in phenylketonuria [J]. Artificial cells blood substitutes and immobilization biotechnology, 1995, 23(1): 1–21.

[108] WANG J F, CHILDERS W S. The future potential of biosensors to investigate the gut-brain axis [J]. Frontiers in bioengineering and biotechnology, 2022, 9: 1–9.

[109] KIMURA H, MIURA S, SHIGEMATSU T, et al. Increased nitric oxide production and inducible nitric oxide synthase activity in colonic mucosa of patients with active ulcerative colitis and crohn's disease [J]. Digestive diseases and sciences, 1997, 42(5): 1047–1054.

[110] ARCHER E J, ROBINSON A B, SUEL G M. Engineered *E. coli* that detect and respond to gut inflammation through nitric oxide sensing [J]. ACS synthetic biology, 2012, 1(10): 451–457.

[111] CHEN X J, WANG B J, THOMPSON I P, et al. Rational design and characterization of nitric oxide biosensors in *E. coli* Nissle 1917 and mini simcells [J]. ACS synthetic biology, 2021, 10(10): 2566–2578.

[112] ROTHHAMMER V, MASCANFRONI I D, BUNSE L, et al. Type I interferons and microbial metabolites of tryptophan modulate astrocyte activity and central nervous system inflammation via the aryl hydrocarbon receptor [J]. Nature medicine, 2016,

22(6):586-597.

[113] WANG J F,ZHANG C,CHILDERS W S. A biosensor for detection of indole metabolites[J]. ACS synthetic biology,2021,10(7):1605-1614.

[114] ZMORA N,ZILBERMAN-SCHAPIRA G,SUEZ J,et al. Personalized gut mucosal colonization resistance to empiric probiotics is associated with unique host and microbiome features[J]. Cell,2018,174(6):1388-1405.

[115] CHARBONNEAU M R,CHARBONNEAU V M,KURTZ C B,et al. Developing a new class of engineered live bacterial therapeutics to treat human diseases[J]. Nature communications,2020,11(1):1738-1749.

[116] RIGLAR D T,GIESSEN T W,BAYM M,et al. Engineered bacteria can function in the mammalian gut long-term as live diagnostics of inflammation[J]. Nature biotechnology,2017,35(7):653-658.

[117] RUBENS J R,SELVAGGIO G,LU T K. Synthetic mixed-signal computation in living cells[J]. Nature communications,2016,7:1-10.

[118] SAGADEVAN A,HWANG K C,SU M D. Singlet oxygen-mediated selective C-H bond hydroperoxidation of ethereal hydrocarbons[J]. Nature communications,2017,8:1-11.

[119] DIMAS R P,JORDAN B R,JIANG X L,et al. Engineering DNA recognition and allosteric response properties of TetR family proteins by using a module-swapping strategy[J]. Nucleic acids research,2019,47(16):8913-8925.

[120] GROSECLOSE T M,RONDON R E,HERDE Z D,et al. Engineered systems of inducible anti-repressors for the next generation of biological programming[J]. Nature communications,2020,11:1-15.

[121] ROTTINGHAUS A G,FERREIRO A,FISHBEIN S R S,et al. Genetically stable CRISPR-based kill switches for engineered microbes[J]. Nature communications,2022,13:1-17.

[122] PAPPOLLA M A,PERRY G,FANG X,et al. Indoles as essential mediators in the gut-brain axis. Their role in Alzheimer's disease[J]. Neurobiology of disease,2021,156:1-11.

[123] KAN A,GELFAT I,EMANI S,et al. Plasmid vectors for in vivo selection-free use with the probiotic *E. coli* Nissle 1917[J]. ACS synthetic biology,2021,10(1):

94–106.

[124] FEDOREC A J H, OZDEMIR T, DOSHI A, et al. Two new plasmid post-segregational killing mechanisms for the implementation of synthetic gene networks in *Escherichia coli*[J]. Iscience, 2019, 14:323–334.

[125] CAMPELO A B, ROCES C, MOHEDANO M L, et al. A bacteriocin gene cluster able to enhance plasmid maintenance in *Lactococcus lactis* [J]. Microbial cell factories, 2014, 13:1–9.

[126] GURBATRI C R, LIA I, VINCENT R, et al. Engineered probiotics for local tumor delivery of checkpoint blockade nanobodies[J]. Science translational medicine, 2020, 12(530):1–12.

[127] PARK Y, BORUJENI A E, GOROCHOWSKI T E, et al. Precision design of stable genetic circuits carried in highly-insulated *E. coli* genomic landing pads[J]. Molecular systems biology, 2020, 16(8):1–19.

[128] GENG P, LEONARD S P, MISHLER D M, et al. Synthetic genome defenses against selfish dna elements stabilize engineered bacteria against evolutionary failure [J]. ACS synthetic biology, 2019, 8(3):521–531.

[129] CALLES J, JUSTICE I, BRINKLEY D, et al. Fail-safe genetic codes designed to intrinsically contain engineered organisms[J]. Nucleic acids research, 2019, 47(19):10439–10451.

[130] SZYDLO K, IGNATOVA Z, GOROCHOWSKI T E. Improving the robustness of engineered bacteria to nutrient stress using programmed proteolysis [J]. ACS synthetic biology, 2022, 11(3):1049–1059.

[131] CERONI F, BOO A, FURINI S, et al. Burden-driven feedback control of gene expression[J]. Nature methods, 2018, 15(5):387–393.

[132] TYAGI A, KUMAR A, APARNA S V, et al. Synthetic biology: applications in the food sector[J]. Critical reviews in food science and nutrition, 2016, 56(11): 1777–1789.

[133] LIU L, GUAN N Z, LI J H, et al. Development of GRAS strains for nutraceutical production using systems and synthetic biology approaches: advances and prospects[J]. Critical reviews in biotechnology, 2017, 37(2):139–150.

[134] GLASS D S, RIEDEL-KRUSE I H. A synthetic bacterial cell-cell adhesion

toolbox for programming multicellular morphologies and patterns[J]. Cell, 2018, 174(3):649-658.

[135] TIMMIS K, TIMMIS J K, BRUSSOW H, et al. Synthetic consortia of nano-body-coupled and formatted bacteria for prophylaxis and therapy interventions targeting microbiome dysbiosis-associated diseases and co-morbidities[J]. Microbial biotechnology, 2019, 12(1):58-65.

第6章 政策导向及发展目标

2023年5月12日，习近平总书记在石家庄市国际生物医药园规划展馆考察时强调：“生物医药产业是关系国计民生和国家安全的战略性新兴产业。要加强基础研究和科技创新能力建设，把生物医药产业发展的命脉牢牢掌握在我们自己手中。要坚持人民至上、生命至上，研发生产更多适合中国人生命基因传承和身体素质特点的‘中国药’，特别是要加强中医药传承创新发展。”

6.1 肠道菌群的产业化应用

微生物会寄生于人体的各个部位，如皮肤、泌尿生殖道、胃肠道和呼吸道。经过长时间的自然选择，寄生在人体上的微生物与宿主形成了共生关系，他们相互关联，相互作用。同时，这些微生物之间也相互作用，形成了稳定的功能群落和生态系统，且占据不同的生态位，共同创造了一个非常稳定的微生物环境[1]。完整的微生物组在许多方面对胃肠道的发展至关重要，微生物与免疫耐受、黏膜相关免疫系统、上皮和屏障功能相关，在宿主中表现为共生的微生物群提供了诸如免疫调节、病原体排斥、细胞保护基因上调、细胞凋亡调节预防和屏障功能维持等稳态功能。

肠道的健康取决于肠道内所有微生物的平衡。也就是说，当肠道微生物群落紊乱时，外源性病原体可以

"趁机进入"宿主，并在肠道内定植和繁殖，从而引发炎症[2]。许多疾病都伴有类似的炎症，临床上通常会给有炎症的患者服用抗生素。后来，研究发现抗生素会对患者的身体造成一些损害，同时，一些患者表现出对抗生素的耐药性。因此，研究者们提出了"微生物群靶向疗法"，包括粪菌移植疗法和益生菌疗法。

6.1.1 粪菌移植疗法

（1）粪菌移植概述

粪菌移植的方法可以追溯到4世纪，当时著名的中国药剂师葛洪用粪便作为药物治疗食物中毒或严重腹泻的患者。16世纪，明朝著名医学家李时珍详细描述了粪便药物的制造方法，并将其命名为"黄龙汤"，用于治疗腹泻、腹痛、呕吐和便秘[3]。直到现代，这种疗法才开始引起人们的注意，并称之为"粪菌移植""粪菌治疗"或"肠道微生物移植"。该方法是指将健康个体的肠道微生物注入患者的肠道，以治疗特定疾病。在治疗中，健康人是供体，而患者是受试者。越来越多的临床试验报告指出：来自健康人的粪便微生物可用作治疗和恢复某些疾病的有效途径，如可实现免疫调节、微生物调节和代谢调节[4]（图6-1）。

大量研究报告称，粪菌移植可以修复艰难梭菌感染引起的肠道微生物紊乱和伪膜性结肠炎[5]。艰难梭菌是一种潜在的病原体，主要与频繁摄入抗生素引起的腹泻有关。艰难梭菌引起的感染会引发重大的健康问题，被称为艰难梭菌感染。研究人员对肠道微生物在艰难梭菌感染发病机制中的作用进行了分析[6]，发现患有复发性艰难梭菌感染的患者表现出肠道微生物组成的变化，且这些变化与频繁服用抗生素有关。在接受粪菌移植的艰难梭菌感染患者身上进行的一项研究中，在粪菌移植前的粪便样本中观察到厚壁菌门和拟杆菌门减少，变形杆菌门增加[7]。另一项针对艰难梭菌感染患者的研究显示，与内毒素产生相关的条件致病菌和产生乳酸的菌群减少（图6-2）。与健康对照组相比，产生丁酸的厌氧细菌增加[8]。这说明粪菌移植能够有效解决由艰难梭菌感染产生的肠道问题。

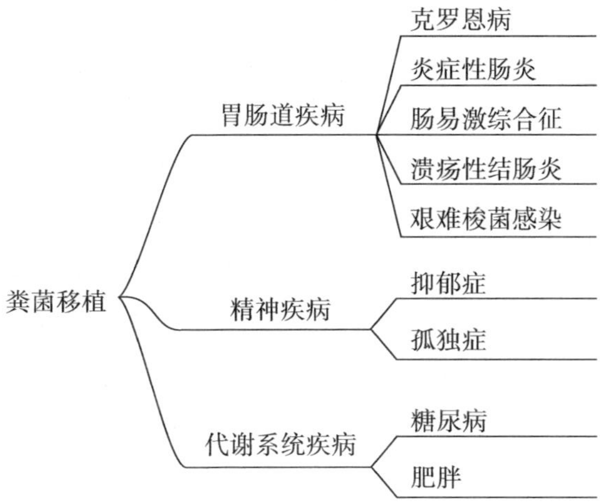

图6-1 利用粪菌移植手段对常见疾病进行治疗[9]

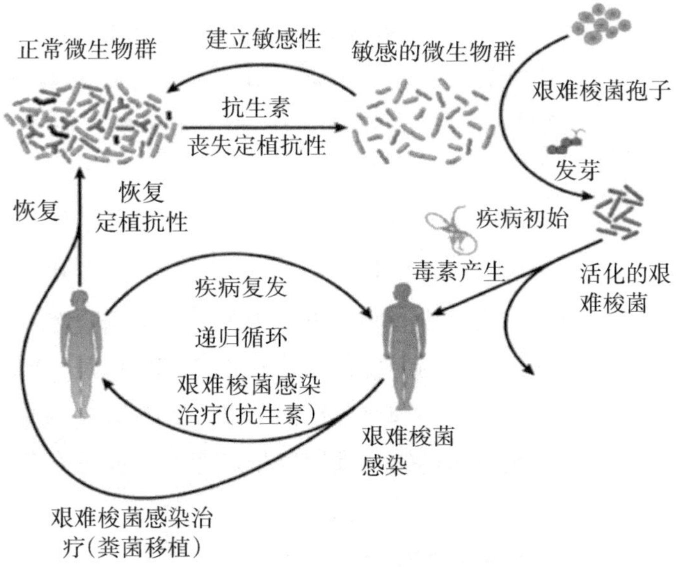

图6-2 艰难梭菌感染治疗与肠道菌群[9]

(2) 粪菌移植方式

粪菌移植可经上消化道、中消化道或下消化道将健康供体的粪便物质转移，常见的移植途径包括鼻胃/鼻腔肠管、胶囊给药、食道胃十二指肠镜、结肠镜、灌肠、经内镜肠植管术、经皮内窥镜盲肠造口术[10]。近些年也有通过口服粪菌制剂治疗溃疡性结肠炎的研究[11]。但是，选择哪种菌液移植方式要根据患者个体情况、病变部位以及疾病的特点制定合适的方案。虽然粪菌移植会有一些不适症状的发生，但都是暂时性的。例如，一项针对在同济医院接受粪菌移植的8547例患者的研究中，随访发现不同的给药方式会造成一定的不适症。鼻胃管组最常见的不良事件是胃管相关的呼吸困难和恶心呕吐。在口服胶囊组中，最常见的不良事件是恶心和呕吐，因为胶囊里有难以完全去除的粪便气味。结肠镜检查组和灌肠组最常见的不良事件是腹泻和肛周不适。而在长期随访和短期随访中，粪菌移植都是较安全的[12]。

随着研究的深入，为了减少粪菌移植的不适症，相关部门先后出台了一些粪菌移植的标准。根据上海市《医疗技术临床应用管理办法》实施细则，上海市卫生健康委员会组织专家研究制定了《上海市菌群移植技术管理规范（2021年版）》，来规范菌群移植技术的临床应用，保证医疗质量和安全。该管理规范的内容包括对医疗机构、人员、技术管理的基本要求以及对培训管理的要求。2022年10月12日，国家市场监督管理总局和国家标准化管理委员会联合发布公告，由张发明教授牵头，李潘等参与制定的国家标准GB/T 41910—2022《洗涤粪菌质量控制和粪菌样本分级》获批执行。这是菌群移植领域发布的第一项国家技术标准，也是洗涤菌群移植的第一项国家技术标准，对于粪菌移植的技术发展和规范实践具有重要意义。该标准明确规定了洗涤粪菌需要的实验室条件和技术要求，确立了洗涤菌群移植的剂量单位。更重要的是，该标准还根据实验室制备方法和安全性等级，明确规定了粪菌样本的安全性等级。

(3) 粪菌移植的局限性与发展展望

粪菌移植是与肠道疾病相关的某些疾病的有效治疗方法，且其治疗成本低。然而，移植整个微生物群落有其自身的风险，一方面，病原体/病理生物的转移、移植会对现有微生物组产生一些不利影响[13]。比如，在粪菌移植期间，复杂的菌群被注入肠道，会对现有肠道微生物组造成实质性破坏。重要的是，引入外来的微生物有时会打破肠道微生物平衡或恶化现有的不适症状[12]。当外来微生物被引入具有独特遗传、免疫、代谢和营养环境的新宿主时，其长期稳定性可能有限。另一个挑战是需确定特定条件下的最佳粪菌移植供体，因为供体选择越来越被认为会对粪菌移植结果产生重大影响[14]

目前也有将工程菌与粪菌移植相结合的研究，工程益生菌能够解决一些粪菌移植的问题，且可作为靶向治疗、预防和诊断的药物。工程益生菌可以在确定感染的地方释放治疗剂，实时监测感染位置的炎症[15]，它通常比普通益生菌有更好的免疫力。因此，应用工程益生菌可以提高治疗效率。

考虑到粪菌移植目前存在的一些不足和面临的一些挑战，在未来应用中，粪菌移植应向改善粪菌提取和移植流程、根据受试者设计个性化的微生物群落、使用特定菌群的靶向治疗等方向发展，从而减轻相关疗法带来的风险。另外，在粪菌移植时，还可以加入益生元辅助治疗，恢复紊乱的肠道菌群及其代谢产物，提高病原菌群的定植力，进而提高粪菌移植的治疗效果。此外，未来还需对粪菌移植制定实施标准，建立标准化粪菌移植流程。

6.1.2 益生菌疗法

益生菌的潜在应用始终涉及保持宿主理想的肠道健康、治疗/预防宿主复发性炎症和免疫系统相关疾病[16]。益生菌在预防和治疗肠道微生物群失调（如急性感染性腹泻和抗生素相关性腹泻）和其他胃肠道疾病（如肠绞痛或肠易激综合征）中具有广泛的应用。在使用益生菌治疗疾病时，患者肠道微生物群落组成变得更稳定，这种改变与疾病症状的恢复呈正相关[17]。同时，益生菌能增强宿主胃肠道的营养和微生物平衡。除此之外，益生菌作为一种载体，能将其有益的功能成分运送到胃肠道的不同目标位置，实现靶向治疗。而活菌菌株的摄入，则具有更有好的治疗效果，但具体情况因菌株而异[18]。

（1）常见益生菌

1）乳杆菌

乳杆菌属涉及各种革兰氏阳性兼性缺氧或微需氧细菌，是乳酸菌群的重要组成部分，主要包括链球菌、明串珠菌、乳球菌、片球菌、双歧杆菌等，这些菌可以将己糖转化为乳酸，从而降低环境的pH值，抑制有害物种的生长[19]。乳杆菌存在于胃肠道和阴道中，其中，双歧杆菌是分娩后最早在婴儿肠道中定植的细菌之一[20]。据报道，乳杆菌属菌株直接影响肠道微生物群，可增强肠道黏膜屏障的功能，有助于维持免疫反应，抑制病原体穿过黏膜屏障，并可用于治疗炎症性肠病、胃肠道感染和肠易激综合征等疾病[21]。

2）凝结芽孢杆菌

芽孢杆菌属为带有芽孢的革兰氏阳性菌，它们要么是严格需氧菌，要么是兼性需氧菌。凝结芽孢杆菌、枯草芽孢杆菌和蜡状芽孢杆菌的孢子在口服时可以抵抗身体的酸性环境，因此适合作为益生菌，并可将其应用于治疗幽门螺杆菌引起的感染和腹泻。单独使用凝结芽孢杆菌或将其与其他微生物（双歧杆菌）联合使用，在治疗由抗生素引起的腹泻方面取得了一定成效[22]。

用凝结芽孢杆菌去除肠道病原体涉及多种机制。

一种机制是在肠道中形成不适合致病性微生物生存的酸性和缺氧条件，从而阻碍致病菌群发育并有助于有益菌群的生长[23]。这是由于凝结芽孢杆菌的兼性性质可使凝固酶菌株在胃肠中消耗游离氧，并减少病原体生长所需的氧化还原反应。而这种环境却适合有益微生物的生长，如双歧杆菌和乳杆菌[24]，从而达到治疗的目的。

另一种机制是在治疗中使用凝结芽孢杆菌时，可以通过产生有助于维持正常微生物群之间平衡的抗微生物物质来抑制病原体的生长[23]。菌株分泌物是其中的一种抗微生物物质，包括几种凝固酶菌株产生的杆菌肽。据报道，凝固酶菌株I4可以产生凝血素，这是一种细菌素样抑制物质（bacteriocin-like inhibitory substances，BLIS）。细菌素是一种阴离子化合物，以不同的方式对抗革兰氏阳性菌，且参与受污染食物引起的各种疾病的治疗[25]。同时，细菌素还具有穿透病原体表面的能力，并可导致无机盐和氨基酸从细胞中泄漏，防止有害细菌的生长[22]。此外，凝固酶菌株还可以分泌乙酸和乳酸，它们是重要的抗微生物物质，同样可以阻止肠道中有害细菌的生长。

3）双歧杆菌

双歧杆菌属是正常人体肠道中构成革兰氏阳性菌群的主要菌属，是一种非能动厌氧菌，可以使阴道和胃肠道之间形成内共生关系。由于益生菌的大部分功能发生在胆盐存在的情况下，因此抗胆盐是其发挥作用的一个重要因素。而双歧杆菌属具有抗胆盐的能力，因此，已被确定为潜在的益生菌。一般而言，野生型乳酸杆菌和双歧杆菌菌株的继代培养以及随后胆汁浓度的增加可导致菌株胆汁耐受性的产生，从而使菌株表现出一定的抗胆盐能力。已报道的双歧杆菌属的益生菌菌株有婴儿双歧杆菌、短双歧杆菌、长双歧杆菌、两歧双歧杆菌、动物双歧杆菌和乳双歧杆菌等[26]。这些双歧杆菌在作为益生菌进行各种疾病的治疗时，表现出的有效性已被广泛证明，如治疗便秘和抗生素引起的腹泻，缓解结肠炎症和结肠中度溃疡，维持肠道稳定等方面。这些菌株还参与预防和治疗食物过敏、辐射引起的腹泻、新生儿坏死性小肠结肠炎、湿疹和高胆固醇等[27]。

（2）天然益生菌疗法

肠道微生物在各种生物功能的调节中都发挥着重要作用，如能量调节、提高宿主对有害微生物的免疫力，以及毒素的中和。研究人员提供了选择特定细菌菌株作为益生菌的不同标准。其中，第一个标准是安全标准，包括益生菌菌株的来源及其非致病性，归类于美国食品和药物管理局（Food and Drug Administration，FDA）评价食品添加剂安全性指标（generally recognized as safe，GRAS）微生物名单。第二个标准涉及益生菌菌株的功能，如抵抗胃肠道酸和胆汁盐的能力、黏附肠道上皮组织表面的能力、调节免疫反应的能力、胃肠道定植及影响人体代谢的能力。第三个标准是技术标准，涉及益生菌的大规模生产，以及益生菌对大规模生产中涉及的各种工艺的抵抗力。总结这些标准，选择益生菌菌株所需的最低标准包括：①益生菌微生物的属和种规格；②必须拥有可行的益生菌种类；③在保质期结束之前，应控制其在适当剂量下具有有益效果（批次之间略有偏差）；④已建立人类对照研究，且证明其有效性和安全性[28]。

在益生菌的临床应用中，对于急性和慢性难治性溃疡性结肠炎患者已经进行了几项益生菌的临床试验。患者经历了复原性全结直肠切除+回肠储袋肛管吻合术（ileal pouch-anal anastomosis，IPAA）的恢复性直肠切除术后，有一定程度的肠胃炎，肠胃炎是回肠储层的一种炎症状态[29]。通过益生菌治疗后，发现益生菌在持续减少炎症方面有很好的安全性和有效性[30, 31]。一些生理学、流行病学和临床研究数据表明，肠道微生物与肠易激综合征的发病机制有关，然而肠易激综合征的病理生理学仍未被发现[32，33]。一项功能研究表明，改变与益生菌结合的宿主肠道微生物可以影响部分宿主肠道功能，如敏感性和运动性，这似乎与肠易激综合征发病机制有关[34]。一项临床试验显示，摄入益生菌的患者（35624例）与摄入安慰剂的患者相比，摄入益生菌的患者疾病症状有更明显的改善。此外，患者血清IL-10/IL12正常化，表明益生菌有助于缓解与肠易激综合征相关的炎症状态[35，36]。此外，植物乳杆菌在缓解肠易激综合征患者的少数症状方面优于安慰剂。例如，DSM 9843菌株从根本上减少了胀气，299 V和LPO1菌株明显降低了肠道疼痛[37-39]。

一项关于在老年患者（65岁以上）和成人患者（18～64岁）中使用益生菌治疗抗生素相关性腹泻的系统综述研究评估了30项符合先前制定的纳入标准的随机管理测试。临床研究表明，益生菌可降低成人中抗生素相关性腹泻的发病率，但在老年人中则不然[40]。研究证明，许多益生菌菌株（如布拉氏杆菌和鼠李糖乳杆菌GG）参与了抗生素相关性腹泻的预防，但其他菌株（如保加利亚乳杆菌、德氏乳杆菌和唾液链球菌）无法预防抗生素相关性腹泻[41-43]。

目前获得专利的主要益生菌包括乳酸菌（如乳酸杆菌、嗜酸乳杆菌、植物乳杆菌、脆皮乳杆菌、罗伊氏乳杆菌、鼠李糖乳杆菌、加斯里乳杆菌、保加利亚乳杆菌）、双歧杆菌（如长双歧杆菌、短双歧杆菌、连环双歧杆菌和动物乳杆菌）和布拉氏杆菌。芽孢杆菌（枯草杆菌、凝固酶、晚孢菌）和粪肠球菌等微生物的潜在用途也已被研究。最常见的商业菌株以乳杆菌属和双歧杆菌属的菌株为主，其相关的益生菌有助于调节和增强动物和人类的先天性免疫和适应性免疫反应（表6-1）。

表6-1　乳酸杆菌和双歧杆菌对不同动物模型先天性和适应性免疫反应的影响

细菌菌株	疾病模型	疾病	功能	参考文献
嗜酸乳杆菌	8周龄雄性C57BL小鼠	炎症性肠病	↑IL-10,调节性T细胞 ↓IL-6、IL-1β、IL-17	[44]
嗜酸乳杆菌（NCK2025）	TS4Cre×APC lox468小鼠	结直肠癌	↑IL-10、IL-12 ↓调节性T细胞	[45]
嗜酸乳杆菌	雌性BALB/c小鼠	克罗恩病	↑IL-17 ↓T17功能、IL-23	[46]
嗜酸乳杆菌	BALB/c小鼠	溃疡性结肠炎	↑乳酸杆菌、双歧杆菌 ↓金黄色葡萄球菌	[47]
干酪乳杆菌BL23	雌性C57BL/6小鼠	结直肠癌	↑T17、T22、IL-10和IL-22 ↓调节性T细胞	[48]
发酵乳杆菌FTDC 812	8周龄BALB/c小鼠	高胆固醇血症	↑乳酸杆菌	[49]
鼠李糖乳杆菌、比菲德氏菌	8周龄C57BL/小鼠	2型糖尿病	↑厚壁菌门、放线菌属 ↓拟杆菌属	[50]
矮型比菲德氏菌IPLA20004	人结肠	炎症性疾病	↑IL-8、IL-10、IL-12	[49]

用作益生菌的细菌、酵母和真菌种类很多。益生菌产生的各种类型的抑制性化合物显示出对致病菌株的拮抗作用。同时，还能产生多种不同的细菌素，如尼克酸，是抗菌作用的主要作用物。例如，乳酸杆菌和双歧杆菌可产生细菌素、乳杆菌素、嗜酸菌素、嗜酸乳杆菌素和乳酸菌素，以防止感染食源性病原体。对这些物种的鉴定可能有助于研究益生菌之间的相互作用以及益生菌的益处。益生菌也可能会增加胃肠道的微生物数量和种类，从而改善肠道菌群的平衡，也可用于治疗各种胃肠道疾病，如肠易激综合征、克罗恩病、大肠杆菌病、抗生素相关性腹泻。除此之外，益生菌还可通过改善肠道微生物群来增强机体免疫，对免疫调节、改善肠道微生物群

和治疗胃肠道疾病至关重要。

（3）工程菌疗法

近年来，大量研究通过现代生物工程技术设计出具有多功能、高效且适应性强的工程菌，主要采用基因编辑策略对微生物特定代谢通路中的关键基因进行过表达或敲除，从而实现对微生物代谢路径的改造，使其分泌有益的代谢产物。已改造的工程菌包括乳酸乳球菌、李斯特菌、大肠杆菌、双歧杆菌等，工程菌在疾病治疗方面取得了一些进展：

①工程益生菌通过动态调节肠道代谢抑制艰难梭菌感染。艰难梭菌感染导致住院患者产生显著的发病率和死亡率，其发病机制本质上与艰难梭菌萌发和孢子形成过程有关，即在活跃的营养细胞和休眠的内共生孢子之间来回移动的能力。通过改造益生菌，即在益生菌中设计一个包含基因编码传感器、放大器和执行器的基因电路，可以恢复肠道胆汁盐代谢，达到抑制由艰难梭菌引起的炎症反应的目的[50]。

②通过口服递送工程益生菌调节肠道菌群和治疗炎症性肠病[51]。炎症性肠病是一种可以诱发结肠和小肠长期炎症的疾病，包括溃疡性结肠炎和克罗恩病，这些疾病可能会进一步诱发更严重的疾病，例如结直肠癌。研究发现使用壳聚糖和海藻酸钠包装的工程益生菌可以修复结肠组织、下调炎症因子表达、显著改变肠道菌群的丰度和多样性。通过基因工程技术对益生菌EcN进行改造，使其过表达过氧化氢酶和超氧化物歧化酶，这两种酶已被证明可有效清除炎症性肠病患者炎症部位的活性氧，故可达到治疗炎症性肠病的目的。

③华东理工大学叶邦策和周英团队构建了一种可用于诊断、记录和缓解炎症性肠病的智能工程益生菌i-ROBOT[52]。i-ROBOT整合了荧光报告、碱基编辑系统和药物分泌等3个核心模块，不仅能实时反映炎症性肠病的

情况并可根据炎症水平释放治疗因子，还能将炎症信息记录到其基因组中。在此研究中，将大肠杆菌Nissle 1917作为底盘，构建了智能工程菌i-ROBOT。i-ROBOT能灵敏地感知炎症标志物硫代硫酸盐，并产生实时信号和可遗传信号，同时可以通过一些反应来监测和记录肠内炎症。另外，i-ROBOT还可根据硫代硫酸盐水平，释放免疫调节蛋白AvCystatin，以改善炎症。

尽管工程菌具有很强的治疗前景，但其临床转化仍有许多局限性。从药物递送的角度看，细菌的固有特性使其成为一种有前景的递送系统，可将治疗药物递送到疾病部位。然而，细菌递送系统的临床转化受限于其潜在毒性和免疫原性，需要进行化学和生物工程化改造[53]。许多细菌具有一定的致病性和强大的定植能力。此外，大多数细菌具有免疫原性，并可能产生副作用，例如引起全身炎症反应综合征，并增加给药后患败血症的风险[54]。此外，细菌很容易被胃酸等恶劣的生理环境破坏，因此口服时会失去治疗效果[55]。工程化改造策略的目的在于去除毒性成分、提高细菌的体内活力和增殖能力、赋予细菌额外的功能，进而促进细菌递送系统的临床转化。

6.2 肠道菌群应用的局限性

目前，益生菌产品的应用越来越多。常见的益生菌服用方式为将其添加到新鲜液体食品中饮用，如酸奶和口服液中[56]。然而，在这种服用方式下能够到达肠道有效部位的活细菌是微不足道的，通常低于最低建议浓度107 cfu/g[57, 58]。究其原因是多方面的，包括：①益生菌对应激反应的抵抗能力是有限的，因为它们中的大多数不会产生孢子，其在生产、运输和储存时就已经死亡；②上消化道中存在的高酸度的胆汁是消化道的天然屏障，可杀死经消化道进入人体的大多数活微生物[58-63]。因此，开发保护活益生菌的生物强化技术迫在眉睫。

6.2.1 肠道菌群的递送和封装技术

微胶囊是一种天然或合成的聚合物包封材料，包封直径为3～800 mm，微胶囊技术是目前常用的益生菌生物强化技术。通常，微胶囊是由半渗透或密封的胶囊膜构成。大量证据表明，微胶囊有效增强了微生物对环境压力（如高温、干燥、胃酸和胆汁）的抵抗力，从而提高了益生菌的稳定性[59，61，64-67]。此外，微胶囊可以保护益生菌免受剪切应力，为微生物生存提供更好的微环境，并有助于提高产品的浓度。总之，当微胶囊形成时，微生物被包壁材料覆盖，可以更好地保持其修复活性。同时，在适当的条件下包壁材料被破坏，益生菌被释放，进一步发挥了其防治功能。微胶囊技术的优点包括：①微胶囊改变了微生态制剂产品的形状，将益生菌转化为稳定状态的粉末，便于运输和储存；②由于微胶囊的保护，可以有效防止细菌的灭活，提高了微生态环境的稳定性；③包壁材料可以保护益生菌不受胃肠道内低pH的影响，能够将益生菌运输到肠道的目标区域；④非水溶性包壁材料可以均匀地转化溶解水中的微生物。因此，微胶囊化有望提高益生菌在生产、储存和消费过程中的稳定性，并可促使生产出耐储存、耐高温、耐高压和耐酸的微生态制剂。

（1）*海藻酸盐基微水凝胶*

海藻酸盐基微水凝胶由于其高生物性能、安全性和低成本特点，已成为最常用的封装载体[61，67-72]。研究表明，在低温和60 ℃时使用海藻酸钠溶液和$CaCl_2$溶液微囊化益生菌，其活菌数增加。然而，人们对藻酸盐的作用仍有争议。尽管一些研究报告了藻酸盐包封增强了乳酸菌在模拟胃肠道条件下的存活率[72-76]，但仍有一些研究发现，将益生菌包封在藻酸盐珠中并不能有效保护微生物免受高酸度的影响[77]。

（2）*淀粉基微水凝胶*

淀粉，包括（微）多孔淀粉，也被应用于微胶囊中

（专利：CN104388416A和CN101904420A）。研究表明，乳酸菌微胶囊与微孔淀粉的包埋率达到90%以上。同时，它增强了在常规水溶液、蒸馏水中的储存稳定性（专利：CN104388416A）。

（3）组合墙材料

单一类型的包壁材料存在一些缺点。例如，海藻酸钠凝胶是多孔的，对影响化合物释放和保护的极端pH值敏感。因此，更多的研究倾向于使用组合型材料进行封装。在含有壳聚糖的藻酸盐珠、含有抗性淀粉和壳聚糖的海藻酸盐珠、壳聚糖包衣的海藻酸盐胶囊、海藻酸盐包衣的明胶微球、海藻酸与淀粉的组合、卡拉胶-海藻酸盐珠和海藻酸-柠檬酸-果胶基质中，细菌细胞的存活率高于纯海藻酸珠[57，76，78-80]。

（4）其他非水溶性墙体材料

随着技术的发展，越来越多的包壁材料被用于生产高性能微胶囊，例如琼脂糖、果胶铁、沸石、合成树脂胶囊、沸石、胞外多糖（exopolysaccharides，EPS）和脂肪蜡[56，65，81-83]。此外，更多具有生物强化作用的微胶囊化技术也处于被开发的阶段，这将扩大益生菌在肠道生物修复中的应用。

6.2.2 肠道菌群作用的靶向性

目前认为菌群与宿主的相互作用具有稳定性和可塑性，稳定性指已经定植的固有菌群对外来菌种具有的定植抵抗性，而可塑性则指肠道微生物的组成和功能存在的个体差异性[84，85]。研究称，非人类谱系的菌群在临床治疗时没有效果[86]，还有研究称，宿主适应的乳酸杆菌菌株在其各自的宿主中显示出更高的生态适应性[87，88]。根据标准指南，益生菌菌株的分离来源必须与其预期的作用部位相关。例如，如果旨在减轻肠道生态失调，那么菌群的来源应该是人类胃肠道。同样，如果打算治疗细菌性阴道病或牙周炎，则必须分别使用阴道或口服分离物分别配制相应的益生菌[87]。因此，肠道菌群在作用时会有一定的靶向性，即从肠道分离的菌群通常作用于肠道，随后通过调节肠道内的代谢物水平，达到治疗疾病的目的。然而，肠道菌群不能直接作用于需要治疗的部位，缺乏治疗靶向性。

6.2.3 肠道菌群作用的个体化差异性

人体肠道微生物群是生活在人体肠道中的微生物，这是一个复杂的微生物群落，估计包含200万亿个细胞，1000多种不同的微生物。人体肠道微生物群由广泛的细菌、真菌、古菌和病毒组成[89]。无论是在肠黏膜表面还是在肠腔内，肠道微生物群均指生活在人类消化道中的微生物。肠道菌群会因个体差异表现出一定的差异。个体一生都有自己稳定的粪便微生物群，并拥有不同的肠道微生物群的特征模式。当机体内引入外来菌群时，初始肠道内的共生微生物种群的殖民抵抗力会使微生物治疗效果变差。此外，微生物组成和功能的个体间差异以及菌株水平差异[84, 85]也可能导致对益生菌[90]或饮食干预[91]的反应性差异，从而限制了菌株治疗的效果。

一项研究报告了乳杆菌FSL-04和罗伊氏乳杆菌ATCC PTA 6475的定植潜力，这两种菌株对人类胃肠道黏膜具有本土性，而嗜酸乳杆菌DDS-1对人类胃肠道具有异质性[92]。该研究得出的结论是，本土菌株在人类胃肠道中的建立比异体菌株更有效。然而，当停止食用时，这两种菌株在第8天都无法被检测到[92]，这突出了益生菌仅在肠道中短暂存在的事实。益生菌菌株在人类胃肠道中定植的潜力取决于宿主微生物群。此外，研究表明，由于个体差异，即使是一个人的本土菌株也可能无法定植于另一个个体的胃肠道[93]。因此，在应用肠道菌群时，个体差异会使治疗效果因人而异。

（胡鑫玉　胡梓剑　毛春兰*）

参考文献

[1]WOODHOUSE C A, PATEL V C, SINGANAYAGAM A, et al. Review article: the gut microbiome as a therapeutic target in the pathogenesis and treatment of chronic liver disease[J]. Aliment pharmacol ther, 2018, 47(2): 192-202.

[2]GOUGH E, SHAIKH H, MANGES A R. Systematic review of intestinal microbiota transplantation (fecal bacteriotherapy) for recurrent *Clostridium difficile* infection [J]. Clinical infectious diseases, 2011, 53(10): 994-1002.

[3]GORDON J I. Honor thy gut symbionts redux[J]. Science, 2012, 336(6086): 1251-1253.

[4]HARTMANN P, CHEN W C, SCHNABL B. The intestinal microbiome and the leaky gut as therapeutic targets in alcoholic liver disease[J]. Frontiers in physiology, 2012, 3: 1-10.

[5]GUARNER F, MALAGELADA J R. Gut flora in health and disease[J]. Lancet, 2003, 361(9356): 512-519.

[6]LESSA F C, MU Y, BAMBERG W M, et al. Burden of *Clostridium difficile* infection in the United States[J]. New england journal of medicine, 2015, 372(24): 23369-2370.

[7]WEINGARDEN A R, CHEN C, BOBR A, et al. Microbiota transplantation restores normal fecal bile acid composition in recurrent *Clostridium difficile* infection[J]. American journal of physiology - gastrointestinal and liver physiology, 2014, 306(4): 310-319.

[8]ANTHARAM V C, LI E C, ISHMAEL A, et al. Intestinal dysbiosis and depletion of butyrogenic bacteria in *Clostridium difficile* infection and nosocomial diarrhea [J]. Journal of clinical microbiology, 2013, 51(9): 2884-2892.

[9]BRITTON R A, YOUNG V B. Interaction between the intestinal microbiota and host in *Clostridium difficile* colonization resistance[J]. Trends in microbiology, 2012, 20(7): 313-319.

[10]GULATI M, SINGH S K, CORRIE L, et al. Delivery routes for faecal microbiota transplants: available, anticipated and aspired[J]. Pharmacological research, 2020, 159: 1-11.

[11]孙岳婷. 粪菌移植治疗慢性便秘的Meta分析[D]. 济南:山东大学,2019.

[12]TIAN H,ZHANG S,QIN H,et al. Long-term safety of faecal microbiota transplantation for gastrointestinal diseases in China[J]. The lancet gastroenterology & hepatology,2022,7(8):702-703.

[13] ALANG N, KELLY C R. Weight gain after fecal microbiota transplantation [J]. Open forum infectious diseases,2015,2(1):1-2.

[14] MOAYYEDI P, SURETTE M G, KIM P T, et al. Fecal microbiota transplantation induces remission in patients with active ulcerative colitis in a randomized controlled trial[J]. Gastroenterology,2015,149(1):102-109.

[15]SCHLOSS P D, HANDELSMAN J. Status of the microbial census[J]. Microbiology and molecular biology reviews,2004,68(4):686-691.

[16]LIN C S,CHANG C J,LU C C,et al. Impact of the gut microbiota,prebiotics, and probiotics on human health and disease[J]. Biomedical journal,2014,37(5):259-268.

[17] CEAPA C, WOPEREIS H, REZAÏKI L, et al. Influence of fermented milk products,prebiotics and probiotics on microbiota composition and health[J]. Best practice & research clininal gastroenterology,2013,27(1):139-155.

[18]ISLAM S U. Clinical uses of probiotics[J]. Medicine(Baltimore),2016,95(5):1-5.

[19] MAKAROVA K, SLESAREV A, WOLF Y, et al. Comparative genomics of the *lactic acid* bacteria[J]. Proceedings of the national academy of sciences-PNAS, 2006,103(42):15611-15616.

[20] WALKER W A. Initial intestinal colonization in the human infant and immune homeostasis[J]. Annals of nutrition and metabolism,2013,63(Suppl 2):8-15.

[21]BAGAROLLI R A,TOBAR N,OLIVEIRA A G,et al. Probiotics modulate gut microbiota and improve insulin sensitivity in DIO mice[J]. Journal of nutritional biochemistry,2017,50:16-25.

[22]RIAZI S,DOVER S E,CHIKINDAS M L. Mode of action and safety of lactosporin, a novel antimicrobial protein produced by *Bacillus coagulans* ATCC 7050[J]. Journal of applied microbiology,2012,113(3):714-722.

[23]HONDA H,GIBSON G R,FARMER S,et al. Use of a continuous culture fermentation system to investigate the effect of GanedenBC(30)(*Bacillus coagulans* GBI-

30,6086)supplementation on pathogen survival in the human gut microbiota[J]. Anaerobe,2011,17(1):36-42.

[24]ABDHUL K,GANESH M,SHANMUGHAPRIYA S,et al. Bacteriocinogenic potential of a probiotic strain *Bacillus coagulans* [BDU3] from Ngari[J]. International journal of biological macromolecules,2015,79:800-806.

[25]ABHARI K,SHEKARFOROUSH S S,HOSSEINZADEH S,et al. The effects of orally administered Bacillus coagulans and inulin on prevention and progression of rheumatoid arthritis in rats[J]. Food & nutrition research,2016,60:1-8.

[26]FUKUI H,OSHIMA T,TANAKA Y,et al. Effect of probiotic *Bifidobacterium bifidum* G9-1 on the relationship between gut microbiota profile and stress sensitivity in maternally separated rats[J]. Scientific reports,2018,8(1):1-10.

[27]PICARD C,FIORAMONTI J,FRANCOIS A,et al. Review article:bifidobacteria as probiotic agents-physiological effects and clinical benefits[J]. Alimentary pharmacology & therapeutics,2005,22(6):495-512.

[28]LEE S J,BOSE S,SEO J G,et al. The effects of co-administration of probiotics with herbal medicine on obesity,metabolic endotoxemia and dysbiosis:a randomized double-blind controlled clinical trial[J]. Clinical nutrition,2014,33(6):973-981.

[29]MCLAUGHLIN S D,CLARK S K,TEKKIS P P,et al. Review article:restorative proctocolectomy,indications,management of complications and follow-up-a guide for gastroenterologists[J]. Alimentary pharmacology & therapeutics,2008,27(10):895-909.

[30]SHEN J,ZUO Z X,MAO A P. Effect of probiotics on inducing remission and maintaining therapy in ulcerative colitis,crohn's disease,and pouchitis:meta-analysis of randomized controlled trials[J]. Inflammatory bowel diseases,2014,20(1):21-35.

[31]PERSBORN M,GERRITSEN J,WALLON C,et al. The effects of probiotics on barrier function and mucosal pouch microbiota during maintenance treatment for severe pouchitis in patients with ulcerative colitis[J]. Alimentary pharmacology & therapeutics,2013,38(11-12):1406-1407.

[32]RINGEL Y,CARROLL I M. Alterations in the intestinal microbiota and functional bowel symptoms[J]. Gastrointestinal endoscopy clinics of north America,2009,19(1):141-150.

[33]SALONEN A,DE VOS W M,PALVA A. Gastrointestinal microbiota in irritable bowel syndrome:present state and perspectives[J]. Microbiology(Reading),2010,156(Pt 11):3205–3215.

[34]MOAYYEDI P,FORD A C,TALLEY N J,et al. The efficacy of probiotics in the treatment of irritable bowel syndrome:a systematic review[J]. Gut,2010,59(3):325–332.

[35]WHORWELL P J,ALTRINGER L,MOREL J,et al. Efficacy of an encapsulated probiotic *Bifidobacterium infantis* 35624 in women with irritable bowel syndrome[J]. American Journal of gastroenterology,2006,101(7):1581–1590.

[36]O'MAHONY L,MCCARTHY J,KELLY P,et al. *Lactobacillus* and *Bifidobacterium* in irritable bowel syndrome:symptom responses and relationship to cytokine profiles[J]. Gastroenterology,2005,128(3):541–551.

[37]NOBAEK S,JOHANSSON M L,MOLIN G,et al. Alteration of intestinal microflora is associated with reduction in abdominal bloating and pain in patients with irritable bowel syndrome[J]. American journal of gastroenterology,2000,95(5):1231–1238.

[38]MCFARLAND L V,DUBLIN S. Meta-analysis of probiotics for the treatment of irritable bowel syndrome[J]. World journal of gastroenterology,2008,14(17):2650–2661.

[39]CARROLL I M,RINGEL-KULKA T,KEKU T O,et al. Molecular analysis of the luminal-and mucosal-associated intestinal microbiota in diarrhea-predominant irritable bowel syndrome[J]. American journal of physiology-gastrointestinal and liver physiology,2011,301(5):799–807.

[40]JAFARNEJAD S,SHAB-BIDAR S,SPEAKMAN J R,et al. Probiotics reduce the risk of antibiotic-associated diarrhea in adults(18–64 Years)but not the elderly(>65 Years):a meta-analysis[J]. Nutrition in clinical practice,2016,31(4):502–513.

[41]SZAJEWSKA H,KOLODZIEJ M. Systematic review with meta-analysis:*Lactobacillus rhamnosus* GG in the prevention of antibiotic-associated diarrhoea in children and adults[J]. Alimentary pharmacology & therapeutics,2015,42(10):1149–1157.

[42]SZAJEWSKA H,KOŁODZIEJ M. Systematic review with meta-analysis:sac-

charomyces boulardii in the prevention of antibiotic-associated diarrhoea[J]. Alimentary pharmacology & therapeutics, 2015, 42(7): 793–801.

[43] PATRO-GOLAB B, SHAMIR R, SZAJEWSKA H. Yogurt for treating antibiotic-associated diarrhea: systematic review and meta-analysis[J]. Nutrition, 2015, 31(6): 796–800.

[44] PARK J S, CHOI J, JHUN J, et al. *Lactobacillus acidophilus* improves intestinal inflammation in an acute colitis mouse model by regulation of th17 and treg cell balance and fibrosis development[J]. Journal of medicinal food, 2018, 21(3): 215–224.

[45] KHAZAIE K, ZADEH M, KHAN M W, et al. Abating colon cancer polyposis by *Lactobacillus acidophilus* deficient in lipoteichoic acid[J]. Proceedings of the national academy of sciences, 2012, 109(26): 10462–10467.

[46] CHEN L, ZOU Y, PENG J, et al. *Lactobacillus acidophilus* suppresses colitis-associated activation of the IL-23/Th17 axis[J]. Journal of immunology research, 2015, 2015: 1–10.

[47] CHEN L L, ZOU Y Y, LU F G, et al. Efficacy profiles for different concentrations of *Lactobacillus acidophilus* in experimental colitis[J]. World journal of gastroenterology, 2013, 19(32): 5347–5356.

[48] JACOUTON E, CHAIN F, SOKOL H, et al. Probiotic strain *Lactobacillus casei* BL23 prevents colitis-associated colorectal cancer[J]. Frontiers in immunology, 2017, 8: 1–10.

[49] AZAD M A K, SARKER M, LI T, et al. Probiotic species in the modulation of gut microbiota: an overview[J]. Biomed research international, 2018, 2018: 1–8.

[50] KOH E, HWANG I Y, LEE H L, et al. Engineering probiotics to inhibit *Clostridioides difficile* infection by dynamic regulation of intestinal metabolism[J]. Nature communications, 2022, 13(1): 1–13.

[51] ZHOU J, LI M, CHEN Q, et al. Programmable probiotics modulate inflammation and gut microbiota for inflammatory bowel disease treatment after effective oral delivery[J]. Nature communications, 2022, 13(1): 1–14.

[52] ZOU Z P, DU Y, FANG T T, et al. Biomarker-responsive engineered probiotic diagnoses, records, and ameliorates inflammatory bowel disease in mice[J]. Cell host & microbe, 2022, 31(2): 199–212.

[53]LI Z,WANG Y,LIU J,et al. Chemically and biologically engineered bacteria-based delivery systems for emerging diagnosis and advanced therapy[J]. Advanced materials,2021,33(38):1–29.

[54]KIM S M,DEFAZIO J R,HYOJU S K,et al. Fecal microbiota transplant rescues mice from human pathogen mediated sepsis by restoring systemic immunity[J]. Nature communications,2020,11(1):1–13.

[55]CAO Z P,WANG X Y,PANG Y,et al. Biointerfacial self-assembly generates lipid membrane coated bacteria for enhanced oral delivery and treatment[J]. Nature communications,2019,10(1):1–11.

[56]WEINBRECK F,BODNAR I,MARCO M L. Can encapsulation lengthen the shelf-life of probiotic bacteria in dry products?[J]. International journal of food microbiology,2010,136(3):364–367.

[57]ETCHEPARE M D A,RADDATZ G C,DE MORAES FLORES E M,et al. Effect of resistant starch and chitosan on survival of *Lactobacillus acidophilus* microencapsulated with sodium alginate[J]. Lwt-food science and technology,2016,65:511–517.

[58]BROECKX G, VANDENHEUVEL D, CLAES I J J, et al. Drying techniques of probiotic bacteria as an important step towards the development of novel pharmabiotics[J]. International journal of pharmaceutics,2016,505(1–2):303–318.

[59]SOHAIL A,TURNER M S,COOMBES A,et al. Survivability of probiotics encapsulated in alginate gel microbeads using a novel impinging aerosols method[J]. International journal of food microbiology,2011,145(1):162–168.

[60]DOHERTY S B,GEE V L,ROSS R P, et al. Development and characterisation of whey protein micro-beads as potential matrices for probiotic protection[J]. Food hydrocolloids,2011,25:1604–1617.

[61]DOHERTY S B,AUTY M A,STANTON C,et al. Survival of entrapped *Lactobacillus rhamnosus* GG in whey protein micro-beads during simulated ex vivo gastro-intestinal transit[J]. International dairy journal,2012,22:31–43.

[62] DIMITRELLOU D, KANDYLIS P, PETROVIC T, et al. Survival of spray dried microencapsulated *Lactobacillus casei* ATCC 393 in simulated gastrointestinal conditions and fermented milk[J]. Lwt-food science and technology, 2016, 71: 169–174.

[63]HUANG S, CAUTY C, DOLIVET A, et al. Double use of highly concentrated sweet whey to improve the biomass production and viability of spray-dried probiotic bacteria[J]. Journal of functional foods, 2016, 23:453–463.

[64]OLIVEIRA A C, MORETTI T S, BOSCHINI C, et al. Stability of *microencapsulated B. lactis*(BI 01) and *L. acidophilus*(LAC 4) by complex coacervation followed by spray drying[J]. Journal of microencapsulation, 2007, 24(7):673–681.

[65]N JIMÉNEZ-PRANTEDA M L, PONCELET D, ÁDER-MACÍAS M E, et al. Stability of *lactobacilli* encapsulated in various microbial polymers[J]. Journal of bioscience and bioengineering, 2012, 113(2):179–184.

[66]KRASAEKOOPT W, BHANDARI B, DEETH H. Evaluation of encapsulation techniques of probiotics for yoghurt[J]. International dairy journal, 2003, 13:3–13.

[67]SU R, ZHU X L, FAN D D, et al. Encapsulation of probiotic *Bifidobacterium longum* BIOMA 5920 with alginate-human-like collagen and evaluation of survival in simulated gastrointestinal conditions[J]. International journal of biological macromolecules, 2011, 49(5):979–984.

[68] SHAHARUDDIN S, MUHAMAD I I. Microencapsulation of alginate-immobilized bagasse with *Lactobacillus rhamnosus* NRRL 442: enhancement of survivability and thermotolerance[J]. Carbohydrate polymers, 2015, 119:173–181.

[69]SOHAIL A, TURNER M S, COOMBES A, et al. The viability of *Lactobacillus rhamnosus* GG and *Lactobacillus acidophilus* NCFM following double encapsulation in alginate and maltodextrin[J]. Food and bioprocess technology, 2013, 6:2763–2769.

[70]SOUSA S, GOMES A M, PINTADO M M, et al. Characterization of freezing effect upon stability of, probiotic loaded, calcium-alginate microparticles[J]. Food and bioproducts processing, 2015, 93:90–97.

[71]YEUNG T W, ÜÇOK E F, TIANI K A, et al. Microencapsulation in alginate and chitosan microgels to enhance viability of *Bifidobacterium longum* for oral delivery [J]. Frontiers in microbiology, 2016, 7:1–11.

[72]LEE K Y, HEO T R. Survival of *Bifidobacterium longum* immobilized in calcium alginate beads in simulated gastric juices and bile salt solution[J]. Applied and environmental microbiology, 2000, 66(2):869–873.

[73]ADHIKARI K, MUSTAPHA A, GRÜN I U, et al. Viability of microencapsu-

lated Bifidobacteria in set yogurt during refrigerated storage[J]. Journal of dairy science,2000,83(9):1946–1951.

[74]ABBASZADEH S, GANDOMI H, MISAGHI A, et al. The effect of alginate and chitosan concentrations on some properties of chitosan-coated alginate beads and survivability of encapsulated *Lactobacillus rhamnosus* in simulated gastrointestinal conditions and during heat processing[J]. Journal of the science of food and agriculture, 2014,94(11):2210–2216.

[75]MANDAL S, HATI S, PUNIYA A K, et al. Enhancement of survival of alginate-encapsulated *Lactobacillus casei* NCDC 298[J]. Journal of the science of food and agriculture,2014,94(10):1994–2001.

[76]HOMAYOUNI A, AZIZI A, EHSANI M R, et al. Effect of microencapsulation and resistant starch on the probiotic survival and sensory properties of synbiotic ice cream[J]. Food chemistry,2008,111:50–55.

[77]SULTANA K, GODWARD G, REYNOLDS N, et al. Encapsulation of probiotic bacteria with alginate-starch and evaluation of survival in simulated gastrointestinal conditions and in yoghurt[J]. International journal of food microbiology,2000,62(1/2): 47–55.

[78]CHÁVARRI M, MARAÑÓN I, ARES R, et al. Microencapsulation of a probiotic and prebiotic in alginate-chitosan capsules improves survival in simulated gastro-intestinal conditions[J]. International journal of food microbiology, 2010, 142(1/2): 185–189.

[79]NAG A, HAN K S, SINGH H. Microencapsulation of probiotic bacteria using pH-induced gelation of sodium caseinate and gellan gum[J]. International dairy journal,2011,21:247–253.

[80]COGHETTO C C, BRINQUES G B, SIQUEIRA N M, et al. Electrospraying microencapsulation of *Lactobacillus plantarum* enhances cell viability under refrigeration storage and simulated gastric and intestinal fluids[J]. Journal of functional foods, 2016,24:316–326.

[81]GHIBAUDO F, GERBINO E, CAMPO DALL' ORTO V, et al. Pectin-iron capsules: novel system to stabilise and deliver lactic acid bacteria[J]. Journal of functional foods,2017,39:299–305.

[82]TAKEI T, HAMADA S, TERAZONO K, et al. Air drying on superamphiphobic surfaces can reduce damage by organic solvents to microbial cells immobilized in synthetic resin capsules[J]. Process biochemistry, 2017, 54:28–32.

[83]HASSANZADEH A M, MAHMOUD S K, SADRNIA M, et al. Immobilization and microencapsulation of *Lactobacillus caseii* and *Lactobacillus plantarum* using zeolite base and evaluating their viability in gastroesophageal - intestine simulated condition [J]. Ars pharmaceutica(Internet), 2017, 58(4):163–170.

[84]ZHU A, SUNAGAWA S, MENDE D R, et al. Inter-individual differences in the gene content of human gut bacterial species[J]. Genome biology, 2015, 16(1):1–13.

[85]NAYFACH S, RODRIGUEZ-MUELLER B, GARUD N, et al. An integrated metagenomics pipeline for strain profiling reveals novel patterns of bacterial transmission and biogeography[J]. Genome research, 2016, 26(11):1612–1625.

[86]SURYAVANSHI M V, PAUL D, DOIJAD S P, et al. Draft genome sequence of *Lactobacillus plantarum* strains E2C2 and E2C5 isolated from human stool culture [J]. Standards in genomic sciences, 2017, 12:1–9.

[87]OH P L, BENSON A K, PETERSON D A, et al. Diversification of the gut symbiont *Lactobacillus reuteri* as a result of host-driven evolution[J]. Isme journal, 2010, 4(3):377–387.

[88]HILL C, GUARNER F, REID G, et al. The international scientific association for probiotics and prebiotics consensus statement on the scope and appropriate use of the term probiotic [J]. Nature reviews gastroenterology & hepatology, 2014, 11(8): 506–514.

[89]QIN J, LI R, RAES J, et al. A human gut microbial gene catalogue established by metagenomic sequencing[J]. Nature, 2010, 464(7285):59–65.

[90]MALDONADO-GÓMEZ M X, MARTÍNEZ I, BOTTACINI F, et al. Stable engraftment of *Bifidobacterium longum* AH1206 in the human gut depends on individualized features of the resident microbiome[J]. Cell host & microbe, 2016, 20(4):515–526.

[91]ZEEVI D, KOREM T, ZMORA N, et al. Personalized nutrition by prediction of glycemic responses[J]. Cell, 2015, 163(5):1079–1094.

[92]FRESE S A,HUTKINS R,WALTER J. Comparison of the colonization ability of autochthonous and allochthonous strains of *Lactobacilli* in the human gastrointestinal tract[J]. Ai magazine,2012,2:399–409.

[93]SIEZEN R J,VAN HYLCKAMA VLIEG J E T. Genomic diversity and versatility of *Lactobacillus plantarum*, a natural metabolic engineer[J]. Microbial cell factories,2011,10:1–13.